AF497620

ALBEIRO PATIÑO BUILES

ELECTRICITY FOR BEGINNERS

FROM ZERO TO MASTER

ALBEIRO PATIÑO BUILES

ELECTRICITY *FOR* BEGINNERS

FROM ZERO TO MASTER

Collection *Engineering*

XALAMBO
EDITORIAL

Collection *Engineering*
ELECTRICITY FOR BEGINNERS. FROM ZERO TO MASTER.
© Albeiro Patiño Builes, 2023
© Xalambo Editorial, 2023

ISBN: 978-628-95735-3-4

Translation: David Esteban Londoño Patiño
Cover design: David Esteban Londoño Patiño
Layout and Editorial Direction: Albeiro Patiño Builes

Edited in Medellín, Colombia
Xalambo Editorial
www.xalambo.com
Tel.: (+57) 302 827 35 13

1st edition: August 2023

«I didn't do anything by accident, nor were my inventions like that; they came through hard work».

Thomas Alva Edison

TABLE OF CONTENTS

1. SOURCES OF ELECTRICAL ENERGY

2. COMPONENTS OF ELECTRICITY

3. ELECTRIC CIRCUITS

4. ELECTRONIC COMPONENTS

5. DIGITAL ELECTRONICS

6. MAGNETISM

7. ELECTRIC POWER SYSTEMS

8. ELECTRICAL SYSTEM CONTROL

9. MEASUREMENT INSTRUMENTS

10. ELECTRICAL SAFETY

INDEX OF FIGURES

1. SOURCES OF ELECTRICAL ENERGY

2. COMPONENTS OF ELECTRICITY

3. ELECTRIC CIRCUITS

4. ELECTRONIC COMPONENTS

5. DIGITAL ELECTRONICS

6. MAGNETISM

7. ELECTRIC POWER SYSTEMS

8. ELECTRICAL SYSTEM CONTROL

9. MEASUREMENT INSTRUMENTS

1. SOURCES OF ELECTRICAL ENERGY

Introduction

Let's imagine for a moment a world without electricity. It would surely be very different from the one we know today, as this resource is essential in many aspects of our daily life.

To begin with, without electricity, there would be no artificial light, which means we would have to live according to the natural rhythms of day and night. On the other hand, the industry and production would be vastly different, as most manufacturing processes rely on the phenomenon of electricity. In other words, the production of food, clothing, electronic devices, and many other consumer goods would be much slower and costlier, or perhaps non-existent. There would also be a significant impact on productivity, as many activities can only be done in daylight. Additionally, without electricity, we wouldn't be able to use the devices we take for granted in our daily lives, such as mobile phones, computers, televisions, and household appliances. This would considerably change the way we communicate, work, and entertain ourselves. Finally, without electricity, many of the medical and technological advancements we have achieved in recent decades would be unusable, affecting healthcare and people's quality of life.

But what is electricity, and where does it come from? Well, electricity is simply a form of energy that is produced by the movement of electric charges, such as electrons, through a conductor. To make these electric charges move, a large number of components linked to a primary source are required.

Sources of electrical energy refer to natural resources, technologies, and systems in general that allow for the generation and supply of electricity through various processes, such as converting energy in all its forms (potential, kinetic, mechanical, chemical, etc.) into electricity. These energy sources can be renewable or non-renewable and are used to produce electricity that powers all kinds of devices, from small household appliances to large industrial infrastructures.

Renewable energy sources

A renewable energy source is one that comes from natural resources that can be replenished or regenerated within a relatively short period of time, such as hydroelectric, solar, wind, thermal, and geothermal energy. These resources are not depleted with continuous use and do not generate significant emissions of greenhouse gases or other atmospheric pollutants. Additionally, their production cost can be lower than that of non-renewable energy sources.

Hydropower

This energy source involves the use of moving water to generate electricity. Water is accumulated in a reservoir and released in a stream that hits the blades of a turbine, which transfer its energy to an electric generator. Figure 1-1 outlines this process.

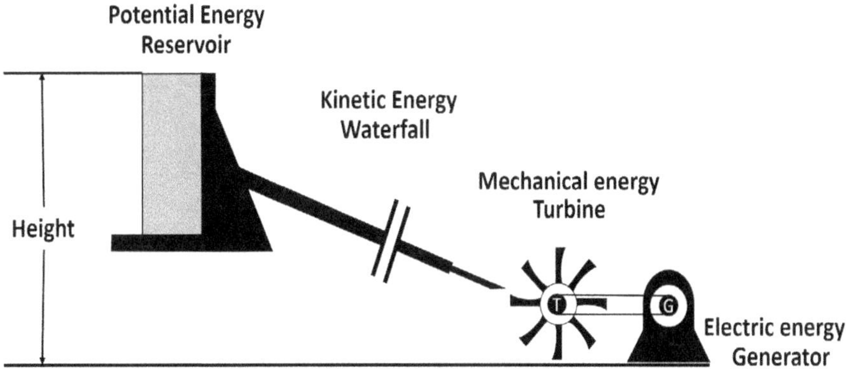

Figure 1-1. Representation of a hydroelectric power plant.

Hydropower is a clean form of generating electricity, but it has a negative impact on the environment as it alters river ecosystems, resulting in a reduction of biodiversity. Often, communities need to be relocated from their original places. Additionally, it is a source of energy that relies on rainfall.

As mentioned before, generating hydropower requires a multitude of components and systems (each designed with specific characteristics and built to fulfill a specific function), but the main ones are as follows:

Turbine

An electric turbine is a device that converts the kinetic energy of a fluid into mechanical energy. It can also operate using the kinetic energy of steam or air. Figure 1-2 shows a graphical representation of this component:

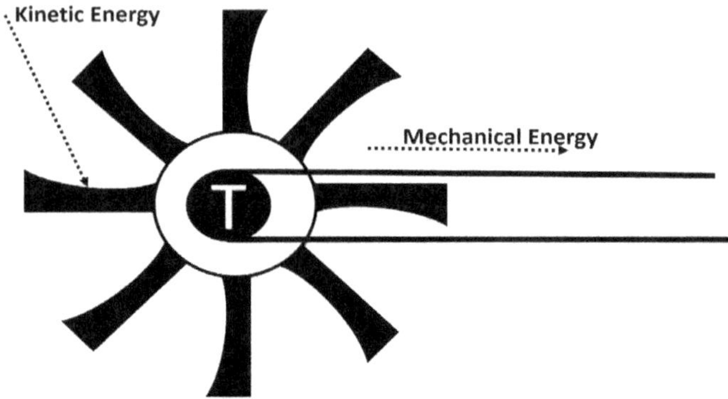

Figure 1-2. Representation of a turbine.

We can mention two fundamental types of turbines: hydraulic turbines, which utilize the kinetic energy of water to generate mechanical energy; this is achieved when water is directed through the blades of the equipment, causing it to rotate. On the other hand, there are steam turbines, which use steam generated by the combustion of fossil fuels or through thermal, geothermal, or nuclear energy to rotate the turbine blades. The movement of the blades, in turn, generates mechanical energy, which is used in the next phase of the generation process.

The mechanical energy produced in the turbines is transferred to a generator, which converts it into electrical energy. Figure 1-3 illustrates the coupling between the shafts of a turbine and an electric generator.

Generator

An electric generator is a device that converts mechanical energy into electrical energy. It is based on the principle of electromagnetic induction, also known as Faraday's law (which we will also cover in this book), to generate electricity.

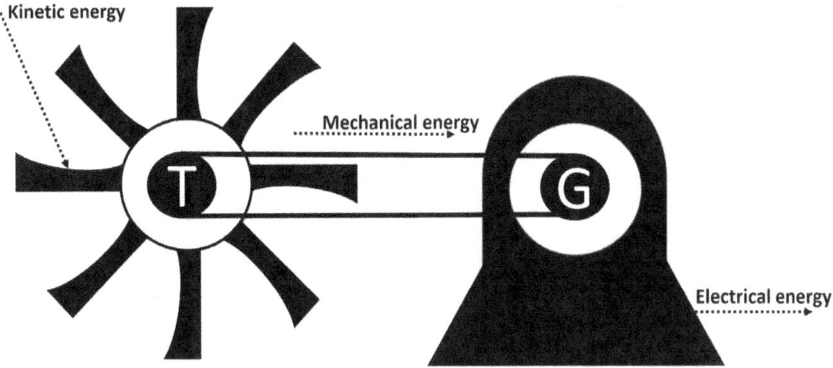

Figure 1-3. Coupling of a turbine and a generator.

The generator consists of three main parts: the rotor, the stator, and a control system, as can be seen in Figure 1-4. The rotor, which is the rotating component, consists of a shaft and a set of magnets or coils. The stator is the stationary component and also consists of a set of windings. When the rotor rotates within the stator, thanks to the mechanical energy transferred by the turbine, its magnets or coils produce a magnetic field that induces an electric current in the windings of the stator.

The control system is responsible for regulating the speed and output power of the generator. If the rotor speed is too high or too low, the electrical output of the generator can be unstable. That's why the control system, which adjusts the rotor speed and output power according to the specific needs of the system, is indispensable.

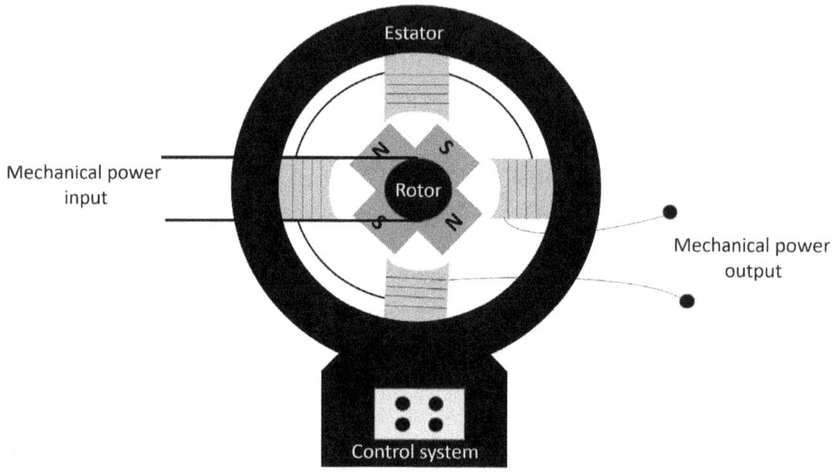

Figure 1-4. Representation of an electric generator.

There are different types of electric generators, but the most common ones are alternating current (AC) generators and direct current (DC) generators. AC generators are the most common and are used in the majority of power plants. DC generators are used in special applications such as motors and some emergency power supply systems.

Solar Energy

Solar energy is a form of renewable energy obtained from the radiation of the sun. It can be used directly, as in the case of heating, or it can be converted into electrical energy through solar panels.

The panels are composed of photovoltaic cells, constructed from semiconductor materials such as silicon, and connected in series to produce the voltage and current required to power electrical devices. Figure 1-5 shows a solar panel with its photovoltaic cells arranged to capture solar energy.

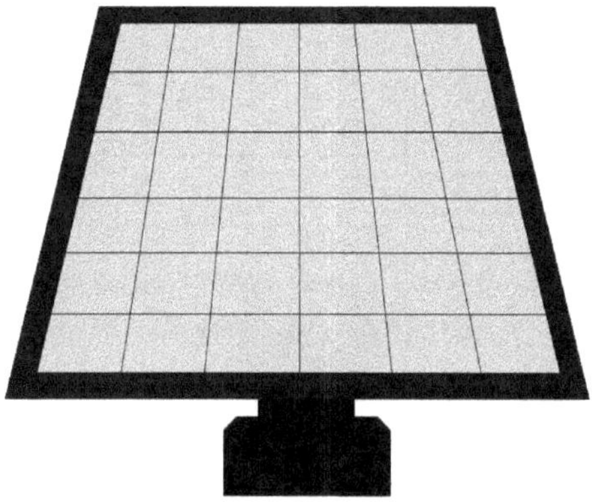

Figure 1-5. Solar panels.

Solar energy has several advantages over other energy sources. Firstly, it is clean, meaning it does not emit greenhouse gases or pollute the environment, and it is renewable, making it an alternative to fossil fuels. Secondly, it is abundant and available worldwide, especially in regions with high levels of solar radiation. Thirdly, solar panels are becoming increasingly affordable and efficient, making them accessible to homes and businesses.

Despite its advantages, however, solar energy also presents challenges. One of the main challenges is the variability of solar radiation, which can limit energy production on cloudy days or during the night. Additionally, the installation and maintenance of these systems may require a significant upfront and ongoing investment, although the costs of this technology are certainly decreasing more and more.

Wind Energy

Wind energy is produced by utilizing wind turbines, which are devices that transform the kinetic energy of the wind into electrical energy.

A wind turbine consists of a tower and one or more blades that rotate when faced with the force of the wind. Figure 1-6 shows an image of a wind turbine. The rotation of the blades drives the generator, which converts mechanical energy into electrical energy.

Wind energy has several advantages. Firstly, it is clean, like solar energy, as it does not emit greenhouse gases or other pollutants, and it is also renewable. Secondly, it is an inexhaustible source of energy available worldwide. Thirdly, the cost of implementation has significantly decreased in recent years, making it more competitive than other conventional forms of energy.

However, like solar energy, wind energy also presents challenges. The main challenge is the variability of the wind, which can affect its production. Additionally, wind turbines can have a visual impact on the landscape and generate noise. However, these issues can be mitigated through proper planning and the use of advanced technologies.

Geothermal energy

This energy source utilizes the heat generated by the Earth to produce electricity. The heat is found beneath the surface and is extracted through geothermal wells. It is also used to generate steam, which rotates the turbines, and in turn powers the generator.

Figure 1-6. Wind turbine.

Non-renewable energy sources

These are natural resources that deplete with continuous use, and their regeneration takes millions of years, such as oil, natural gas, and coal. Non-renewable resources are often more expensive to produce and contribute significantly to the emission of greenhouse gases and other atmospheric pollutants, resulting in a negative impact on the environment and human health. Additionally, their supply can be vulnerable to interruptions or price fluctuations due to political or economic factors.

Thermal energy

This energy source involves the burning of fossil fuels such as coal, oil, or natural gas to generate heat and produce steam. This steam is used to spin the turbines of a generator, which in turn produces electricity. Despite being a common form of energy generation, its environmental impact is very negative due to the greenhouse gas emissions it produces.

Nuclear energy

EThis source utilizes the energy released during nuclear fission to generate heat and produce steam. Similar to the previous case, this steam is used to spin turbines that drive a generator. Nuclear energy is a highly potent form of generating electricity, but its environmental impact can be very negative due to the management of radioactive waste and the risk of nuclear accidents.

Review Questionnaire

Answer the question:

1. What is an electrical energy source?

2. How does a hydroelectric power plant work?

3. What is a turbine and what is it used for?

4. How is solar energy generated?

5. What is electrical energy transmission and what is its importance in power supply?

Select the correct option:

6. Which of the following energy sources are renewable?
    a. Nuclear energy
    b. Thermal energy
    c. Wind energy
    d. Geothermal energy

7. Which device is used to convert mechanical energy into electrical energy in a hydroelectric power plant?
    a. Generator
    b. Turbine
    c. Solar panel
    d. Windmill

8. What is the primary energy source used to generate electricity in thermal power plants?
    a. Oil
    b. Natural gas
    c. Coal
    d. Nuclear energy

9. What type of energy is generated from solar radiation?
    a. Hydroelectric energy
    b. Wind energy
    c. Geothermal energy
    d. Solar energy

Indicate if the statement is false or true:

10. Nuclear energy is a renewable energy source.
    a. False
    b. True

11. Wind energy is generated from the wind.
    a. True
    b. False

12. In a hydroelectric power plant, the turbine is responsible for converting electrical energy into mechanical energy.
    a. False
    b. True

13. Solar energy can be stored in batteries for later use.
    a. True
    b. False

14. Electrical energy transmission is the process of generating electrical energy from a power source.
    a. False
    b. True

Complete the statement with the correct word or expression:

15. Wind energy is generated from the force of __________.

16. Hydroelectric energy is produced through the mechanical energy generated by the movement of water in a __________.

17. Thermal energy is generated from the heat produced by the combustion of __________.

18. Solar panels convert solar energy into __________.

2. COMPONENTS OF ELECTRICITY

Introduction

Electrical components are the fundamental building blocks of any circuit. These components are used in most of the electronic devices we use daily, from mobile phones and computers to cars and large-scale electrical power systems. Therefore, knowledge of electrical components is essential for anyone working in the electronics industry, from designers and technicians to electrical engineers and science students. Understanding how they operate and how they are combined to create larger and more complex circuits is key to developing increasingly efficient and advanced technologies.

Electrical components can be divided into two categories: active components and passive components. The former includes batteries, generators, power sources, transistors, and integrated circuits, while the latter comprises resistors, capacitors, inductors, and diodes.

For example, electrical resistance is a fundamental property of materials that is used to vary the flow of current in a circuit. Capacitors and inductors are important for energy storage in the circuit, while batteries and generators are energy sources used to power it.

Let's now take a look at some of the components of electricity.

Charge

Electric charge is a property of subatomic particles (electrons, protons, and neutrons) that gives them the ability to interact with each other through electromagnetic forces. Charge can be positive or negative, measured in Coulombs (C), and represented by the symbol Q. Electrons have a negative charge, while protons have a positive charge. Neutrons have no electric charge or, as their name suggests, are neutral.

Electric charges can be transferred between objects through direct contact or the process known as induction, which we will discuss later. When two objects have opposite electric charges, they attract each other, as shown in Figure 2-1(a), where a positive and a negative charge attract each other. When they have like electric charges, they repel each other, as shown in Figures 2-1(b) and (c), where positive charges repel each other and negative charges repel each other.

Furthermore, electric charges generate electric fields around them (there are also magnetic fields and electromagnetic fields). An electric field is a region of space in which a charge experiences an electric force. The magnitude of the field depends on the quantity and distribution of charges present in the region.

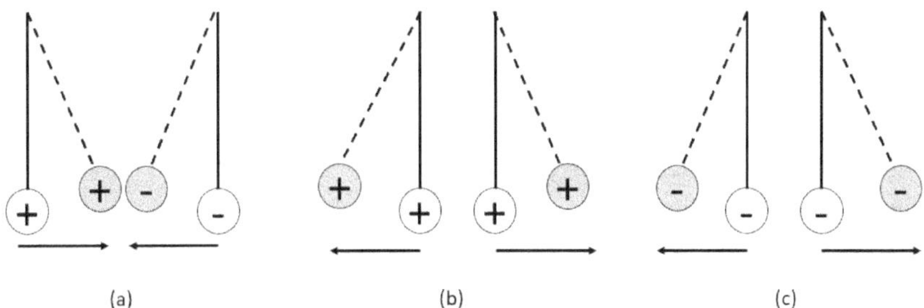

Figure 2-1 (a) Attraction of different charges, (b) Repulsion of positive charges, (c) Repulsion of negative charges.

The force between electric charges is governed by Coulomb's Law, while the force between charges in motion and magnetic fields is governed by Lorentz's Law.

Coulomb's Law

Coulomb's Law states that the electric force between two point charges is directly proportional to the product of their charges and inversely proportional to the square of the distance between them. Mathematically, it can be expressed as follows:

$$F = k \times \left(\frac{q_1 \times q_2}{d^2} \right)$$

Where:

F, is the electric force between the charges
K, is known as the Coulomb constant
q_1, is the charge of particle 1

q_2, is the charge of particle 2

d, is the separation distance between the two charges

Electric charges have numerous applications in everyday life. For example, electricity is a form of energy that relies on the movement of these charges; electronic devices use them to operate, and magnets can generate electric currents through the relative movement of charges in a circuit. Additionally, electric charges are utilized in industries for processes such as electrophoresis (a laboratory technique used to separate molecules of DNA, RNA, or proteins based on their size and charge) or even electric welding.

Current

Electric current is the flow of charges through a conducting material, such as a wire or an electrical circuit, within a certain period of time. Mathematically, this can be expressed as follows:

$$I = \frac{Q}{t}$$

Where:

I, is the electric current

Q, is the electric charge

t, is the time

For there to be an electric current, a potential difference between two points is required. The potential difference creates an electric field that drives the charges through the conducting material, thus generating the current. Current is measured in amperes (A), represented by I, and its flow direction is defined as the direction in which positive charges move.

Electric current can be of two types: direct current (DC) or alternating current (AC), as shown in Figures 2-2(a) and (b) respectively. Direct current flows in a single direction, while alternating current changes direction periodically. Alternating current is the form in which electricity is supplied to most homes and businesses.

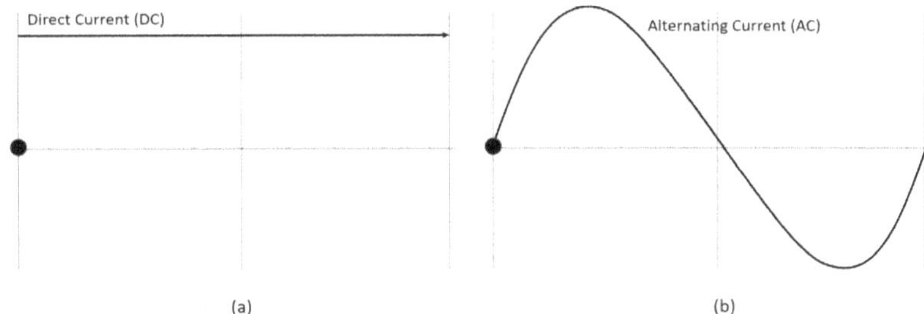

Figure 2-2 (a) Direct Current (DC) and (b) Alternating Current (AC).

Now, if the intensity of the electric current depends on the amount of charge flowing through the conductor per unit of time, increasing the charge will increase the intensity of the current (later on, we will see that this is achieved by increasing the voltage or reducing the resistance in a circuit).

Electric current can be dangerous if not handled properly. If a person touches a live conductor while current is flowing through it, the person can become a conductor of electricity, and the current will flow through their body, which can cause injuries or even death. It is important to exercise caution when working with electricity and follow proper safety practices.

Voltage

Electric voltage (or potential difference or electrical tension) is the measurement of the electric force that drives current through a circuit. It is measured in volts (V), represented by a V, and can be thought of as the pressure that pushes electrons through a conductor. In other words, voltage is created when there is a difference in electric charge between two points in an electrical circuit; if a conductor is connected between those two points, electrons will flow from one to the other, thereby generating the current. Figure 2-3 illustrates an analogy between the difference in water level in a hydraulic system and the potential difference in an electrical circuit.

Note that the height H in the figure represents a force by which the water tries to find equilibrium on both surfaces. Similarly, when there is an excess of electrons on side A of the graph and a deficit on side B, electrons will flow from A to B, trying to fill the gaps in B.

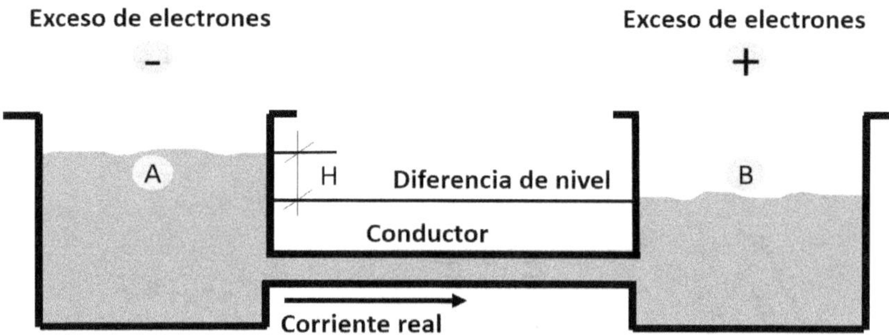

Figure 2-3. Analogy between the water level difference in a hydraulic system and the potential difference in an electrical system.

Voltage is important, for example, to power electrical and

electronic devices, and to charge batteries and accumulators that power various equipment, as well as to transmit electricity over long distances through high-voltage lines.

Electric voltage can be dangerous if not handled properly. If a person touches two cables simultaneously that have a potential difference applied between them, an electric current can flow through their body, which can cause injuries or even death. Caution should be exercised when working with electricity and appropriate safety practices should be followed.

Power

Electric power is the measure of the amount of electrical energy transferred to or consumed by an electrical circuit per unit of time. It is measured in watts (W), represented by the letter P, and is directly proportional to the current and voltage in the circuit. It can be calculated using the formula:

$$P = VI$$

Where:

P, is the electric power in watts
V, is the electric voltage in volts
I, is the electric current in amperes.

In the diagram in Figure 2-4, a power source in a circuit is shown, where in (a) it supplies electric power, while in (b) it absorbs electric power (observe the direction of current flow).

Electric power can be classified into two types: active power and reactive power. Active power is the electrical energy that is converted into useful work, such as the power used to operate a light bulb or a motor. Reactive power is the electrical energy that is stored and released in a circuit, such as the energy accumulated in a capacitor or an inductor.

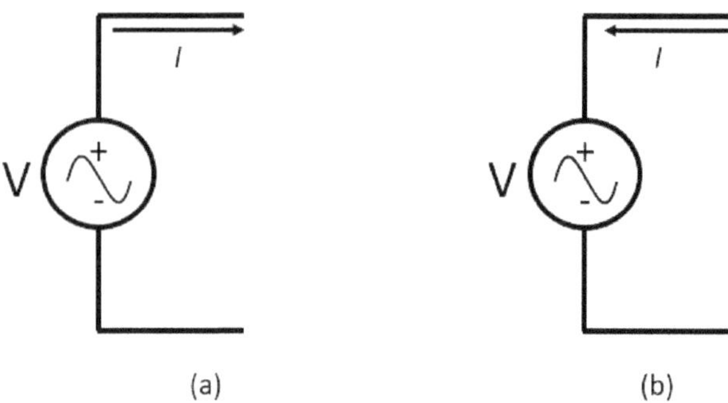

(a)   (b)

Figure 2-4. (a) Power supplied by a source, (b) Power absorbed by a source.

Power is necessary to determine the capacity of a system, to supply electrical and electronic devices, and to design electrical protection systems, such as fuses and circuit breakers, which protect the circuit and its connected devices against electrical overloads. No amount of warnings will be enough for those who work with or use electricity to be cautious of its dangers.

Since electrical power results from current and voltage, it is important to note that it can also be dangerous if caution is not exercised when working with it. It is essential to follow proper safety practices, whether dealing with high voltage and low current electrical circuits or low voltage and low current electronic circuits. Electricians must always ensure that devices are designed and connected according to applicable specifications and standards.

Energy

Electric energy is produced through the transfer of electrons between atoms. It is a highly versatile form of energy used in a wide range of applications, from lighting and operating appliances to powering industrial and transportation equipment worldwide. It is measured in joules (J) or kilowatt-hours (kWh) and represented by the letter E. A kilowatt-hour is the amount of energy consumed when using a device with a power of one kilowatt for one hour.

As we have seen, electric power can be generated from various sources such as hydroelectric, thermal, wind, solar, or nuclear power. The most common sources utilize the force of water or steam to rotate turbines, which in turn drive generators to produce electricity.

Electric energy is transmitted and distributed through a network of power transmission and distribution lines (we will explore these aspects in more detail later). In the distribution network, electric energy is transformed to different voltage levels for use in homes, businesses, and industries.

On the other hand, measures must be taken to reduce unnecessary electricity consumption and promote the use of renewable and sustainable energy sources for its generation. Once again:

Electric energy can be dangerous if not handled properly. It is important to follow appropriate safety practices and adhere to applicable norms and regulations when working with it.

Resistance

Electrical resistance is the property exhibited by materials to oppose the flow of current. It is a measure of the difficulty that current encounters while passing through a material, and it is measured in ohms (Ω), represented by the symbol R. The symbol shown in Figure 2-5 is commonly used to represent resistance.

Figure 2-5. Symbol of electrical resistance.

From another perspective, electrical resistance is due to the interaction of moving electrons (current) with the atoms of the material through which they move. When electrons collide with the atoms of the material and lose energy, it increases the resistance, thereby reducing the flow of electricity.

Mathematically, resistance can be calculated in terms of the voltage applied across it and the current passing through it.

$$R = \frac{V}{I}$$

Where:

R, is the resistance of the material
V, is the applied voltage across the material

I, is the current passing through the material

Later on, we will see that this relationship is known as Ohm's Law, which has significant applications in the design of electrical circuits as it allows the determination of current, voltage, or resistance if the other two values are known.

Electrical resistance can affect the performance and safety of a circuit. For example, the use of cables with excessively high resistance can cause voltage drop and energy loss in the circuit. On the other hand, the use of resistors in series or parallel (configurations we will discuss later in this book) with other components can adjust the flow of current and protect the elements against damage.

Conductors, Semiconductors, and Insulators

Materials with low electrical resistance are called conductors, while those with high electrical resistance are called insulators. Materials with intermediate electrical resistance are called semiconductors. Each of these groups of elements has specific applications in the field of electricity. Table 2-1 provides examples of conductive, insulating, and semiconductor elements.

Table 2-1. Conductors, insulators, and semiconductors

Conductors	Insulators	Semiconductors
Copper	Glass	Silicon
Aluminum	Ceramic	Germanium
Gold	Wood	Selenium
Silver	Air	Tellurium

Iron	Rubber	Gallium
Zinc	Porcelain	Gallium arsenide
Nickel	Plastic	Amorphous silicon
Bronze	Teflon	Silicon carbide

Electric Conductivity of Materials can be explained as follows:

Conductive Materials: In conductive materials, the electrons in the outermost shell of atoms (known as valence electrons) are loosely bound to the nucleus and can easily move between atoms. When an electric field is applied (e.g., by applying a voltage), electrons can move through the material, resulting in an electric current.

Semiconductor Materials: Semiconductor materials have intermediate electrical conductivity between conductors and insulators. Valence electrons in semiconductors are more strongly bound to the nucleus compared to conductors, but they can still move with some ease. However, the electrical conductivity of semiconductors depends heavily on temperature and the presence of impurities in the material.

Insulating Materials: In insulating materials, valence electrons are strongly bound to the nucleus and cannot move easily. Therefore, these materials do not conduct electricity easily.

Electronic Configuration

Electronic configuration refers to the arrangement of electrons in different energy levels and sublevels of an atom or ion.

Electrons are distributed in different orbitals around the nucleus of the atom and are organized into electron shells called energy levels.

The notation used to describe electronic configuration involves the sequence of quantum numbers n, l, and m, which indicate energy, shape, and orientation of the orbital, respectively. "n" represents the principal quantum number, indicating the energy level or shell in which the electrons are located. This value can be any positive integer greater than or equal to 1. "l" represents the secondary or orbital angular momentum quantum number, indicating the shape of the orbital in which the electron is found. This value can be any non-negative integer smaller than n. "m" represents the magnetic quantum number, indicating the orientation of the orbital in space. This value can be any integer within the range of -l to l.

The number of electrons that can occupy a given energy level is determined by the formula $2n^2$, where n is the principal quantum number.

For example, the electronic configuration of oxygen atom (Z=8) is $1s^2\ 2s^2\ 2p^4$, indicating that it has 2 electrons in the 1s energy level, 2 electrons in the 2s level, and 4 electrons in the 2p level. If we add up the electrons in the outermost energy level, i.e., the 2nd level, we find that it has 6 electrons.

Conductors have fewer than 4 valence electrons in their outermost energy level, meaning they are loosely bound to the nucleus and can easily be dislodged to produce an electric current. *Semiconductors* have exactly four valence electrons, and *insulators* have more than four valence electrons.

The electronic configuration of an atom or ion can be used to predict its chemical behavior as well as its physical and chemical properties. Additionally, electronic configuration can be influenced by the presence of other atoms, an applied electric field, temperature, among other factors. Thus, electronic configuration can vary in different situations and is an important property for understanding the chemistry and physics of atoms and molecules.

The topic of electronic configuration of materials is crucial for understanding their electrical behavior, but it is beyond the scope of this book and is left as a research activity.

Coil and Inductance

Electrical inductance is a property of circuits that relates to the ability of a conductor to generate a magnetic field in response to the flow of current passing through it. It is measured in henries (H) and is represented by the letter L. The symbol for inductance can be seen in Figure 2-6.

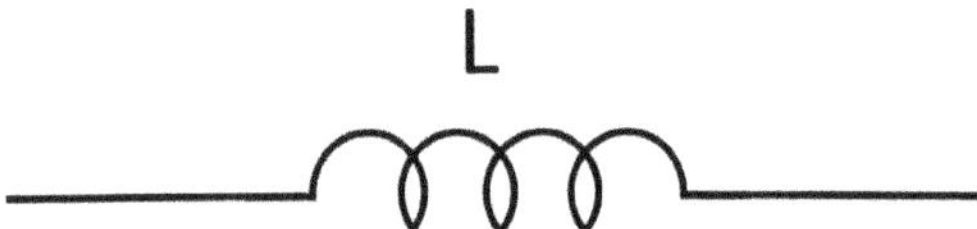

Figure 2-6. Symbol of electrical coil.

When an electric current flows through a conductor, a magnetic field is generated that extends around it. The magnetic field can interact with nearby conductors, thereby inducing an electric current in them. This property is known as electrical inductance and occurs in circuits that contain coils or solenoids. Figure 2-7 illustrates how solenoids are constructed.

Electrical inductance depends on the geometry of the coil, the number of wire turns, the core material, and the magnetic permeability of the material. As the electric current changes over time, the magnetic field around the coil also changes, which can produce an electromotive force in it that counteracts the change in electric current. This electromotive force is known as self-induction.

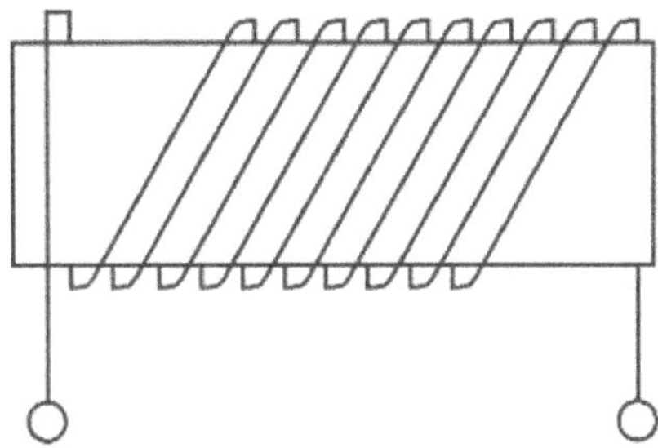

Figure 2-7. Example of a coil or solenoid.

Inductance can also be present in circuits that do not contain coils or solenoids. Any circuit that has a spiral or circular conductor can have electrical inductance.

This component is used in many electronic devices, such as transformers, electric motors, generators, and filtering circuits. Transformers use electrical inductance to transform the alternating current voltage from one level to another, while electric motors use inductance to generate a magnetic field that produces movement in the rotor.

Capacitor and capacitance

Electrical capacitance is a property of electrical circuits that relates to the ability of a capacitor to store electric charge. It is measured in farads (F) and is represented by the letter C.

When a voltage is applied to a capacitor, electric charge accumulates on the opposite surfaces of the component. Capacitance refers to the amount of charge that a capacitor can store in relation to the applied voltage. Figure 2-8 shows the symbol of a capacitor.

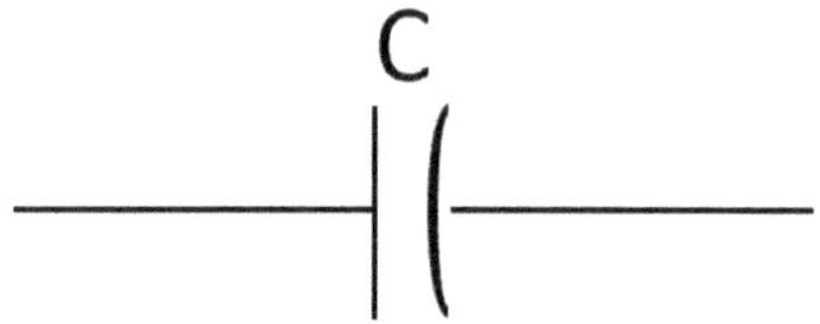

Figure 2-8. Symbol of an electrical capacitor.

Electrical capacitance depends on the geometry of the capacitor, the separation between the conductive surfaces, and the insulating material separating them. The larger the surface area of the capacitor, the higher the electrical capacitance. On the other hand, the larger the separation between the conductive surfaces, the lower the capacitance. Figure 2-9 shows how a capacitor is constructed:

Electrical capacitance is used in many electronic devices, such as filters, oscillators, and coupling circuits. It is also used in circuits to store energy, eliminate electrical noise, and filter signals. It is important in data transmission circuits as the capacitance of a cable can cause a decrease in transmission speed and degradation in signal quality.

Battery

An electrical battery is a device that converts chemical energy into electrical energy through an electrochemical reaction. It consists of one or more electrochemical cells, which are in turn composed of two electrodes made of different materials, an anode and a cathode, separated by an electrolyte (see the Voltaic cell in Figure 2-10). When an electrical circuit is connected to the electrodes, a chemical reaction occurs that releases electrons at the anode and captures them at the cathode, producing an electric current that can be used to power devices.

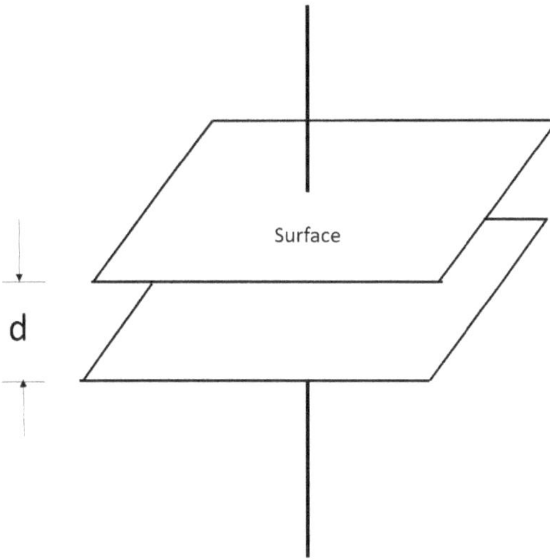

Figure 2-9. Structure of a capacitor.

Some batteries are rechargeable and can be recharged many times, while others are disposable and need to be replaced when depleted. Rechargeable batteries are recharged using an external power source (a charger), which reverses the chemical reaction to return the stored energy to the battery.

Electrical batteries are widely used in portable electronic devices such as mobile phones, music players, laptops, and cameras, as well as in electric vehicles and energy storage systems. They are available in a wide variety of sizes and shapes, from small button batteries used in watches and headphones to large lithium-ion batteries used in electric vehicles and energy storage systems. Each type of battery has its own performance characteristics, capacity, lifespan, and cost.

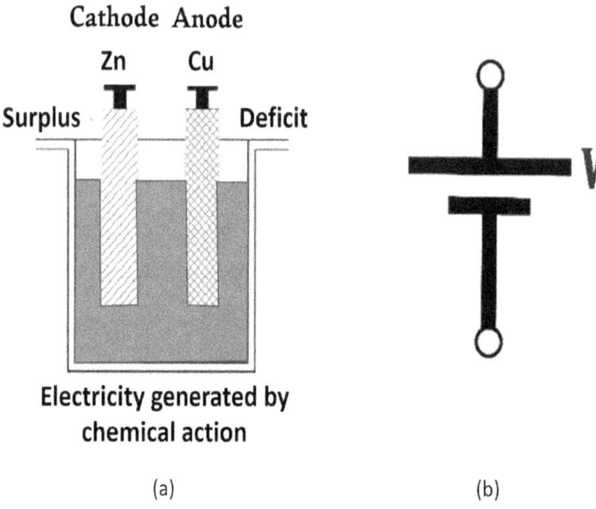

Figure 2-10 (a) Voltaic pile, (b) symbol of a battery.

Generator

An electrical generator is a device that converts mechanical energy into electrical energy. Generators are commonly used to provide electricity in power plants or remote areas. Figure 2-11 shows the symbol of a generator.

Independent source

An independent source of electricity is a device that provides electrical energy autonomously and continuously to an electrical circuit. Unlike dependent power sources, an independent source of electricity supplies a constant voltage regardless of variations in the load.

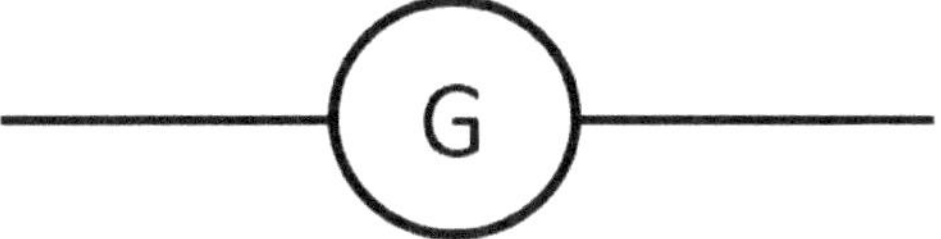

Figure 2-11. Symbol of an electric generator.

Independent sources are often used in applications where stable and reliable voltage is required, such as industrial process control systems, medical equipment, and communication systems.

There are different types of independent sources, including linear power supplies and switched-mode power supplies. The former use a transformer and a voltage regulator to provide a stable and fluctuation-free output to the circuit. The latter use high-frequency switching technology to convert the energy from a variable input voltage source into a constant and regulated output voltage. Figure 2-12 shows the symbols of an independent voltage source (a) and an independent current source (b).

Independent sources of electricity can also be classified according to their current capacity and voltage capacity. Low-power supplies are used in portable electronic devices, while high-power supplies are used in industrial applications.

Dependent source

A dependent source of electricity (also called regulated or controlled) is a device that provides an output that depends on an electrical signal in another component of the circuit. Unlike independent sources, a dependent source adjusts its output in response to a specific input signal.

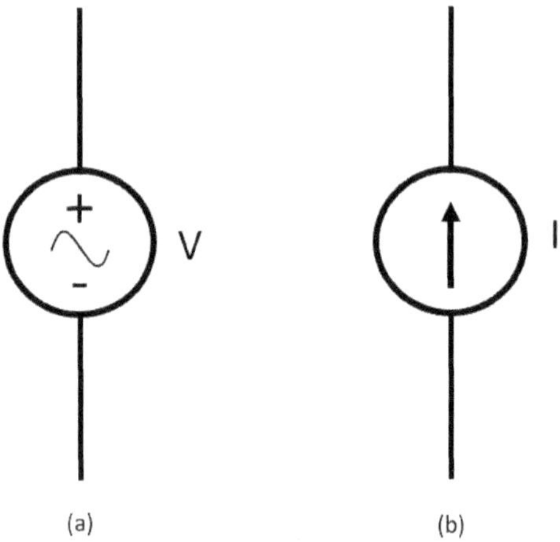

Figure 2-12. Independent source (a) of voltage, (b) of current.

There are different types of dependent sources, including current-controlled sources (Figure 2-13b and d) and voltage-controlled sources (Figure 2-13a and c). Current-controlled sources are used to provide controlled current in response to a specific current signal elsewhere in the circuit, while voltage-controlled sources are used to provide controlled voltage in response to a specific voltage signal elsewhere in the circuit

Dependent sources of electricity can also be classified according to their relationship to the input signal. Proportional sources have a linear relationship between the input signal and the source output, while non-proportional sources have a non-linear relationship between the input signal and the source output

Dependent sources of electricity are often used in signal amplification and control applications, such as audio amplifier circuits, oscillators, and automatic control systems.

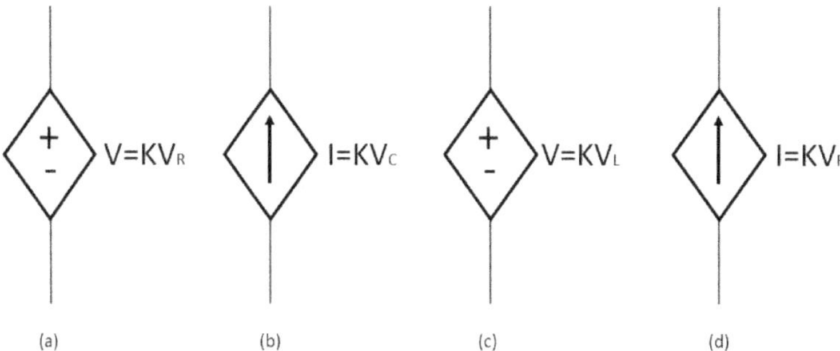

Figure 2-13. Dependent sources of voltage and current.

To conclude this chapter, Table 2-1 summarizes the components reviewed in it, along with their representation letter and the unit of measurement they work with.

Table 2-1. Summary of Electrical Components

Component	Representation	Meaning	Unit
Charge	Q	Amount of charge	Coulomb (C)
Current	I	Current intensity	Ampere (A)
Voltage	V	Voltage difference	Volt (V)
Power	P	Power	Watt (W)
Energy	E	Energy	Joule (J)
Resistance	R	Resistance	Ohm (Ω)
Inductance	L	Inductance	Henry (H)
Capacitance	C	Capacitance	Farad (F)
Battery	B	Stored charge	Coulomb (C)
Generator	G	Generated voltage	Volt (V)

Independent Source	-	Independent voltage or current	Volt (V) or Ampere (A)
Dependent Source	-	Dependent voltage or current	Volt (V) or Ampere (A)

Review questionnaire

Answer the question:

1. What is electrical charge?

2. What is the unit of measurement for electrical charge?

3. What is the definition of electric current?

4. Is electric current dangerous or harmless?

5. Explain in your own words electric voltage.

6. What are other names for electric voltage?

7. What is electrical power?

8. How is electrical power calculated?

9. Is there any difference between a Joule and a kWh?

10. What sources of electrical energy do you know?

11. All elements have a certain electrical resistance. Explain.

12. What is a semiconductor?

13. What is electrical inductance?

14. How would you build an electrical inductance?

15. What is electrical capacitance?

16. What is the unit of measurement for electrical capacitance?

17. Where and for what purpose are electric batteries used?

18. What are electric generators?

19. An independent source of electricity depends only on time. Explain.

20. How many types of controlled sources of electrical energy can you conceive?

3. ELECTRIC CIRCUITS

Definition

An electrical circuit is a network of interconnected electrical components that allows the flow of current. Typical components of a circuit include energy sources such as batteries or generators, as well as devices that consume that energy, such as resistors, coils, and capacitors.

Electric circuits can be designed to fulfill different functions, such as providing power to a device, performing a specific task like a timer, or controlling an automated system. They are also used in communication solutions, consumer electronics, medical equipment, and various industries.

They can be analyzed using Ohm's law and Kirchhoff's laws, as we will see later, which establish the relationships between currents, voltages, and resistances in a circuit.

Types of Electric Circuits

There are several types of electric circuits, which differ in their configuration and the flow of current they handle. Some of the most common types include:

Series Circuits

Where the components are connected one after another, so that the current flows through them sequentially. The current is the same throughout the circuit, but the voltage is divided among the connected elements. If one of these components fails, the entire circuit is interrupted. Figure 3-1 shows an example of a series circuit.

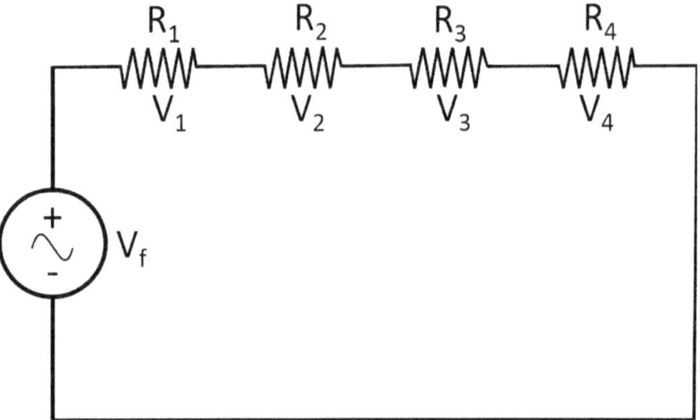

Figure 3-1. Series circuit

Parallel Circuits

Where the components are connected in separate branches, so that the current is divided among them. Each branch may have different values of resistance and current, and if one component fails, the others continue to function. Figure 3-2 illustrates a parallel circuit.

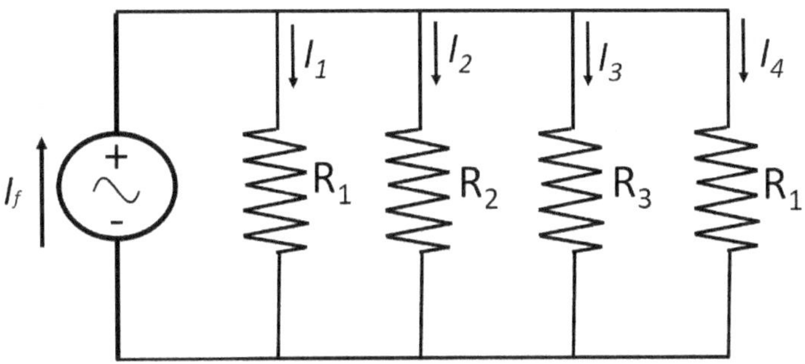

Figure 3-2. Parallel circuit

Mixed Circuits

Combine elements of series and parallel circuits. They are typically divided into separate sections, each with its own parallel branches and series components. Figure 3-3 shows an example of a mixed circuit.

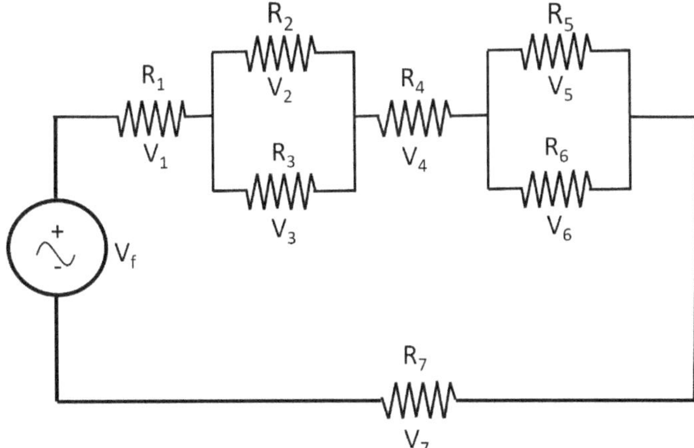

Figure 3-3. Mixed circuit

Direct Current (DC) Circuits

These circuits are designed to be powered by direct current, which, as we know, flows in a single direction. This type of circuit is common in batteries and power supplies.

Alternating Current (AC) Circuits

These circuits are designed to be used with alternating current, which changes direction and magnitude periodically. They are common in electric power transmission and in most lighting systems and electronic devices that use grid electricity.

It is important to note that the types of electric circuits used in a specific application will depend on the requirements of that application. For example, series circuits are suitable for Christmas lights, where all lights need to turn on at the same time or none at all. On the other hand, parallel circuits are useful for wall outlets, where multiple devices can be plugged into different sockets, and each one will function independently.

Ohm's Law

Ohm's law establishes the relationship between current, voltage, and resistance in an electric circuit. This law was discovered by the German physicist Georg Simon Ohm in 1827 and states that:

> *The electric current flowing through a circuit is directly proportional to the applied voltage and inversely proportional to the circuit's resistance.*

The constant that relates current and voltage in Ohm's law is precisely the electrical resistance of the element.

Mathematically, it can be expressed as follows:

$$I = \frac{V}{R}$$

Where:

I, is the electric current in amperes (A)
V, is the voltage in volts (V)
R, is the resistance in ohms (Ω)

This means that if the voltage remains constant, an increase in the circuit's resistance will decrease the electric current flowing through it. Similarly, if the resistance is held constant, an increase in voltage will increase the electric current.

Ohm's law is a fundamental tool for the design and analysis of electric circuits. It allows engineers and technicians to calculate the current, voltage, or resistance necessary for a circuit to function correctly. Moreover, it is used in the measurement and testing of electrical components and systems.

Kirchhoff's Laws

Kirchhoff's laws are two fundamental principles that govern the behavior of electric currents in a closed circuit. The first is Kirchhoff's current law, and the second is Kirchhoff's voltage law. Both are widely used in circuit analysis, from basic electronics to advanced engineering. They were established by the German physicist Gustav Kirchhoff in 1845 and state that:

Kirchhoff's Current Law (KCL)

The sum of the currents entering a node in a circuit is equal to the sum of the currents leaving that node.

This is due to the conservation of electric charge, which cannot be created or destroyed. Mathematically, it can be expressed as follows:

Where ΣI is the algebraic sum of the currents entering and leaving a node.

$$\sum_{1}^{n} I = 0$$

Kirchhoff's Voltage Law (KVL)

The sum of the voltages in a closed circuit is equal to zero.

Similar to the current law, this is because electric energy cannot be created or destroyed. Mathematically, it can be expressed as follows:

Where ΣV is the algebraic sum of the voltages in a closed circuit.

$$\sum_{1}^{n} V = 0$$

Kirchhoff's laws are essential for the analysis of complex electric circuits and are frequently used by electrical professionals to address their design needs. By applying these laws to a circuit, the currents and voltages in different components and parts of the circuit can be determined, allowing for a better understanding of the circuit's behavior and the solution of electrical problems.

Circuit Analysis

Circuit analysis is a discipline of electrical engineering that focuses on the study and resolution of complex circuits. It is essential for the design, construction, and maintenance of electrical and electronic systems, ranging from simple lighting circuits to complex automation and control systems.

Circuit analysis is often performed using mathematical techniques and simulation, where not only nodal analysis and mesh analysis are relevant, but also other techniques such as Laplace transform analysis, Fourier analysis, and computer simulation. These tools enable an understanding of how circuit variables (voltage, current, resistance, capacitance, inductance, etc.) interact with each other and how they affect the performance of the circuit.

Circuit analysis also involves the use of measurement instruments (which will be discussed later in this book), such as multimeters, oscilloscopes, and spectrum analyzers, to make precise measurements of electrical variables in the circuit and verify its real-world functioning.

Nodal Analysis Method

Nodal analysis is a technique for analyzing electrical circuits that is used to find the currents in various components of the circuit. In this method, the circuit is analyzed in terms of voltages at nodes, which are the points where three or more components connect, as shown in Figure 3-4.

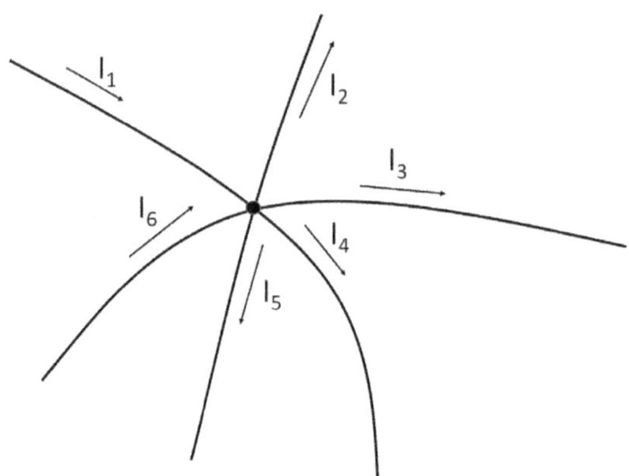

Figure 3-4. Electric node

The process of analysis using the nodal analysis method involves the following steps:

1. Identify the nodes of the circuit and label them with numbers.
2. Select a reference node and assign it a potential of zero volts.
3. Assign voltages to the other nodes and define them in terms of the voltage at the reference node.
4. Apply Kirchhoff's current law at each node in the circuit, which means that the sum of the currents entering and leaving each node must be equal to zero. In the diagram

above, it would be:

$$I_1 + I_6 = I_2 + I_3 + I_4 + I_5$$

5. Solve the equations from the previous step to find the unknown currents in the circuit components.

The nodal analysis method is particularly useful for analyzing complex electrical circuits and is a commonly used tool in electricity-related engineering fields.

Mesh Analysis Method

Mesh analysis is a technique for analyzing electrical circuits that is used to find the currents in various components of the circuit. In this method, the circuit is analyzed in terms of the currents flowing through the different meshes of the circuit, as shown in Figure 3-5.

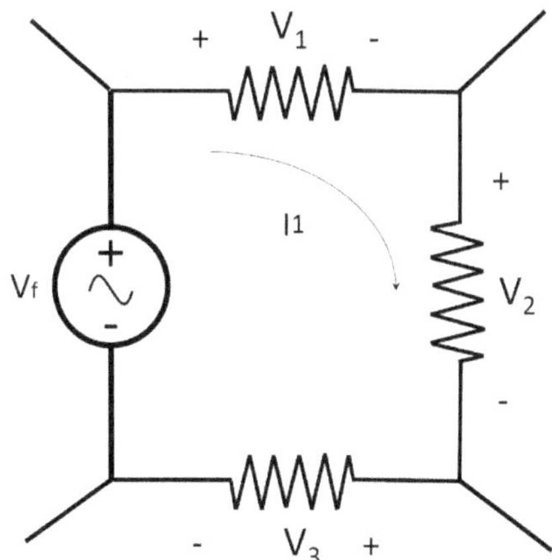

Figure 3-5. Electric mesh

The process of analyzing using the mesh analysis method involves the following steps:

1. Identify the meshes of the circuit and label them with a letter. A mesh is a closed path in the circuit that does not contain interior nodes.
2. Assign unknown currents to the different meshes of the circuit.
3. Apply Kirchhoff's voltage law to each mesh of the circuit, which means that the sum of the voltage drops in each mesh must be equal to zero. In the previous diagram, it would be:

$$V_f = V_1 + V_2 + V_3$$

4. Solve the resulting equations from the previous step to find the voltages and then the unknown currents in the circuit components.

It is important to note that Kirchhoff's voltage law is applied only to the circuit elements that are part of the mesh.

This method is particularly useful for analyzing electrical circuits with multiple voltage and current sources, as well as circuits with nonlinear elements.

Voltage and Current Dividers

Voltage and current dividers are useful tools for controlling and adjusting the amount of voltage or current in a circuit, and their use extends from basic electronics to more complex industrial electronics and communication systems.

In a circuit, a voltage divider divides the voltage from a voltage source in a determined ratio using series resistors. On the other hand, a current divider is a circuit used to divide the current from a current source in a determined ratio. It is similar to a voltage divider, but instead of dividing the voltage, it divides the input current into two or more branches. Let's examine them in more detail below:

Voltage Divider

It is used to divide an input voltage into two or more smaller voltages. It consists of a series of resistors connected in series between the power source and the load. The output voltage is taken from a point in the resistor chain. The relationship between the input voltage and the output voltage is determined by the ratio of the resistance values in the chain. For example, if there are two equal resistors in series, the output voltage at the intermediate point will be half of the input voltage. In the circuit of Figure 3-6, the output voltage of the voltage divider is taken at the connection point between the two resistors.

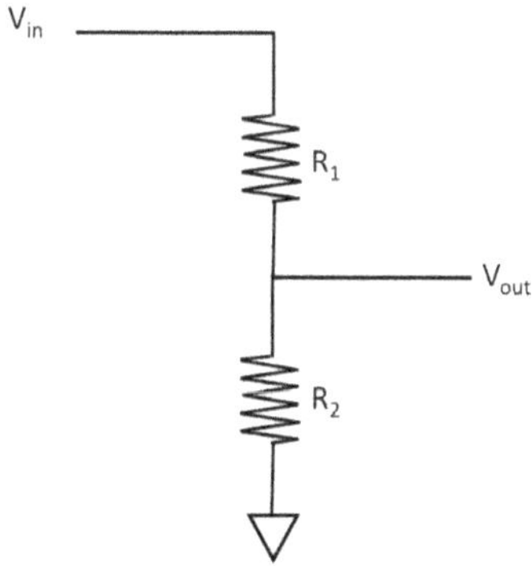

Figure 3-6. Voltage divider

The equation that describes voltage division in this case is:

$$V_{out} = V_{in} \left(\frac{R_2}{R_1 + R_2} \right)$$

Where:

V$_{out}$, is the output voltage
V$_{in}$, is the input voltage
R$_1$, is the resistance connected to the input terminal
R$_2$, is the resistance connected to the output terminal

The ratio between the resistances determines the voltage division ratio. For example, if R1 is equal to R2, the output voltage will be half of the input voltage. If R2 is much smaller than R1, the output voltage will be a smaller fraction of the input voltage.

It is important to note that a voltage divider consumes power, so the current flowing through the circuit can affect its performance. Additionally, the input impedance of the circuit connected to the output of the voltage divider can affect the accuracy of the output voltage. Therefore, it is important to consider these factors when designing and using these circuits.

Current Divider

It is used to divide an input current into two or more smaller currents. It consists of a parallel combination of resistors connected in parallel between the power source and the load. The output current is taken from a point in the resistor chain. The relationship between the input current and the output current is determined by the ratio of the resistance values in the chain. For example, if there are two equal resistors in parallel, the output current through each resistor will be half of the input current.

The basic circuit of a current divider uses two resistors in parallel connected in series with the current source. The input current is divided in a determined proportion, with a fraction of the current flowing through one resistor, and the rest through the other. The total current flowing through both resistors is equal to the input current, as shown in Figure 3-7:

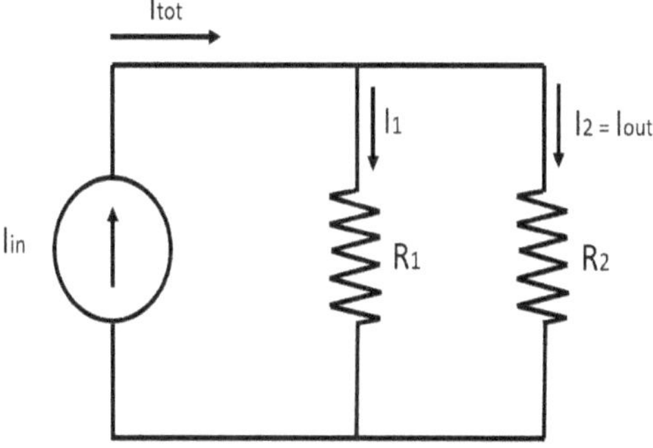

Figure 3-7. Current divider

The equation that describes current division is:

$$I_1 = I_{in}\left(\frac{R_2}{R_1 + R_2}\right)$$

$$I_2 = I_{out} = I_{in}\left(\frac{R_1}{R_1 + R_2}\right)$$

Where:

I_1 and $I_2 = I_{out}$, are the currents flowing through R_1 and R_2, respectively.

I_{in}, is the input current.
R_1 and R_2, are the resistances connected in parallel.

The relationship between the resistances determines the current division ratio. For example, if R1 is equal to R2, the current is divided equally between both branches. If R2 is much larger than R1, most of the current will flow through R1.

Similar to the voltage divider, a current divider consumes power and can affect the accuracy of the output current due to the connected load. Therefore, it is important to consider these factors when designing and using a current divider.

Transformations Δ to Y and Y to Δ

Sometimes, in circuit analysis, we come across combinations that are not easily solvable using the methods of node and mesh analysis. Some of these cases are known as star and delta circuits. They can also be referred to as T and π configurations, or Y and Δ configurations. Both versions are shown in Figure 3-8.

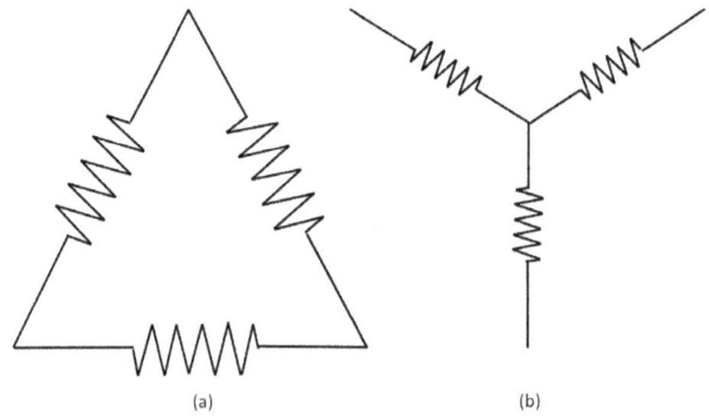

Figure 3-8. Circuit (a) in Δ and (b) in Y

In any case, a Y combination can always be transformed into a Δ, and likewise, a Δ can always be transformed into a Y. However, the reason these types of circuits are also called T and π is that, in circuit practice, they often resemble these shapes, as can be seen in Figure 3-9.

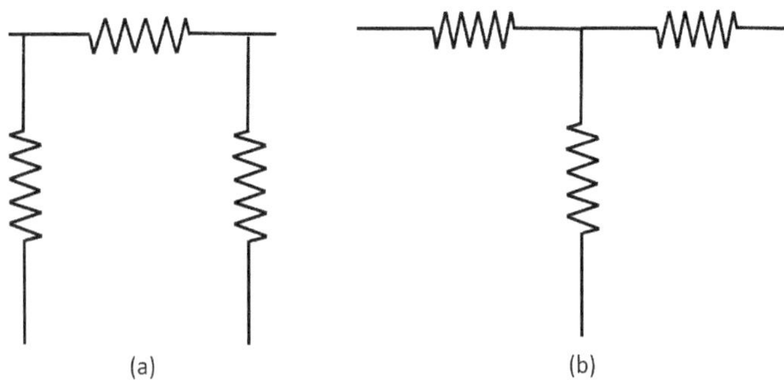

Figure 3-9. Circuit (a) in π and (b) in T

The Y to Δ conversion is commonly used in electrical engineering to connect three impedances in a circuit. If the impedances are connected in a Y configuration, it can be converted to a Δ combination using the following relationship, based on Figure 3-10:

Δ to Y Transformation

The equations to transform a Δ combination into a Y are:

$$R_1 = \frac{R_b R_c}{R_a + R_b + c}$$

$$R_2 = \frac{R_a R_c}{R_a + R_b + c}$$

$$R_3 = \frac{R_a R_b}{R_a + R_b + c}$$

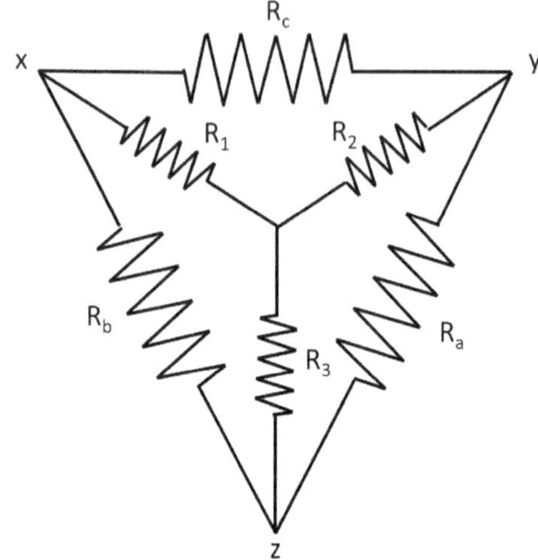

Figure 3-10. Y to Δ and Δ to Y transformation

Where R_a, R_b and R_c are the impedances of the branches in the Δ configuration.

Y to Δ Transformation

On the other hand, if the impedances are connected in a Y configuration, this combination can be converted to a Δ using the following relationship, based on the same Figure 3-10:

Where R_1, R_2 and R_3 are the impedances of the branches in the Y configuration.

The Y to Δ transformation is useful in applications such as power distribution and power electronics (which we will cover later) where it is necessary to connect multiple loads in a balanced three-phase circuit.

$$R_a = \frac{R_1 R_2 + R_2 R_3 + R_3 R_1}{R_1}$$

$$R_b = \frac{R_1 R_2 + R_2 R_3 + R_3 R_1}{R_2}$$

$$R_c = \frac{R_1 R_2 + R_2 R_3 + R_3 R_1}{R_3}$$

Review Questionnaire

Select the correct option:

1. What are three types of electrical circuits?
    a. In series, in parallel, and mixed
    b. In series, in parallel, and combined
    c. In series, in parallel, and alternating

2. What is a series circuit?
    a. A circuit in which all components are connected in sequence
    b. A circuit in which all components are connected in parallel
    c. A circuit in which all components are randomly connected

3. What is a parallel circuit?
    a. A circuit in which all components are connected in a line
    b. A circuit in which all initial ends of the components are connected, and likewise, all final ends of those components are also connected.
    c. A circuit in which all components are randomly connected

4. What is an impedance?
    a. The total opposition that an electrical circuit presents to the flow of alternating current
    b. The total resistance in an electrical circuit
    c. A measure of the amount of current flowing through a circuit

5. What is a direct current circuit?
    a. A circuit in which the current is constant and flows in one direction
    b. A circuit that contains variable current
    c. A circuit that contains alternating current

6. What is an alternating current circuit?
   a. A circuit that contains constant current
   b. A circuit in which the current periodically changes direction
   c. A circuit that contains direct current

7. How can Ohm's law not be expressed?
   a. $V = I/R$
   b. $I = V/R$
   c. $R = V/I$

8. What are Kirchhoff's laws?
   a. Kirchhoff's voltage law and Kirchhoff's current law
   b. Kirchhoff's resistance law and Kirchhoff's capacitance law
   c. Kirchhoff's charge law and Kirchhoff's electric field law

9. What is circuit analysis?
   a. The process of determining and interpreting the behavior of an electrical circuit
   b. The process of connecting electrical components in series
   c. The process of connecting electrical components in parallel

10. What is nodal analysis?
   a. The process of determining the voltage at a specific node in a circuit
   b. The process of determining the currents entering and leaving a specific node in a circuit
   c. The process of determining the resistance at a specific node in a circuit

11. What is mesh analysis?
   a. The process of determining the inductance in the components that make up a specific mesh in a circuit
   b. The process of determining the current flowing through the electrical components in a specific mesh in a circuit
   c. The process of determining the resistance in a mesh in a circuit

Answer the question:

12. What is an electrical circuit?

13. In the circuit in Figure 3-11, determine the value of current i_2, knowing

that i_1 = 10A, i_3 = 3, i_4 = 4A, i_5 = 5A. What can you conclude from the obtained result?

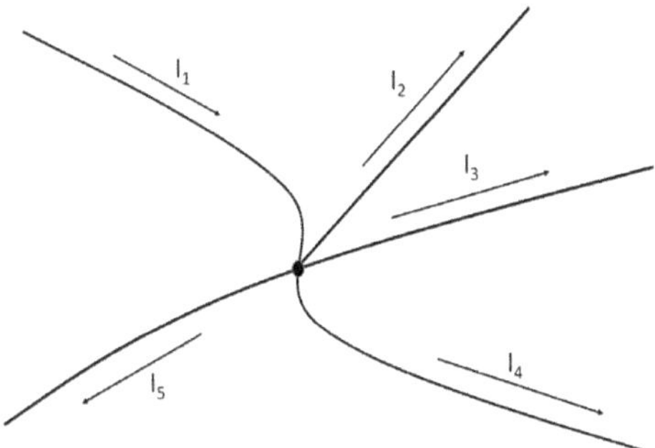

Figure 3-11. Circuit from exercise 3-13

14. In the circuit in Figure 3-12, determine the value of voltage V_1, knowing that V_f =12V, V_2 = 3V, V_3 = 4V, i_4 = 5V. What can you conclude from the obtained result?

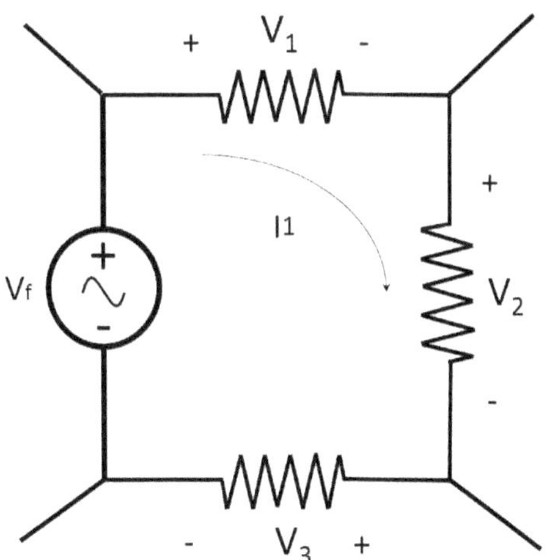

Figure 3-12. Circuit from exercise 3-14

15. In the circuit in Figure 3-13, find the values of the impedances Ra, Rb y Rc, knowing that in (a) R1 = 4 Ω, R2 = 5Ω and R3 = 8Ω.

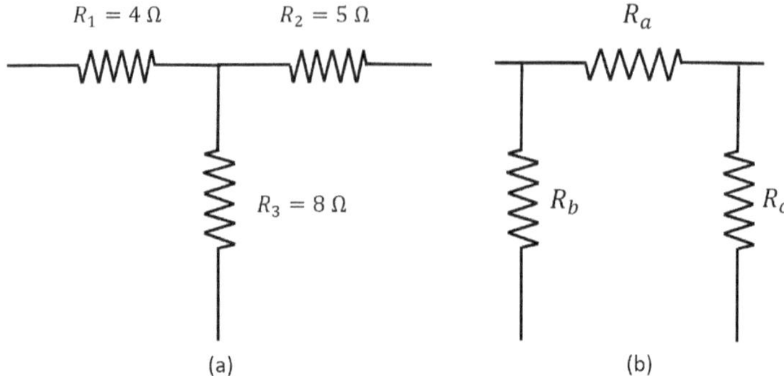

Figure 3-13. Circuit from exercise 3-15

16. In the circuit in Figure 3-14, find the values of the impedances Ra, Rb and Rc, knowing that in (a) R1 = 4 Ω, R2 = 5Ω and R3 = 8Ω.

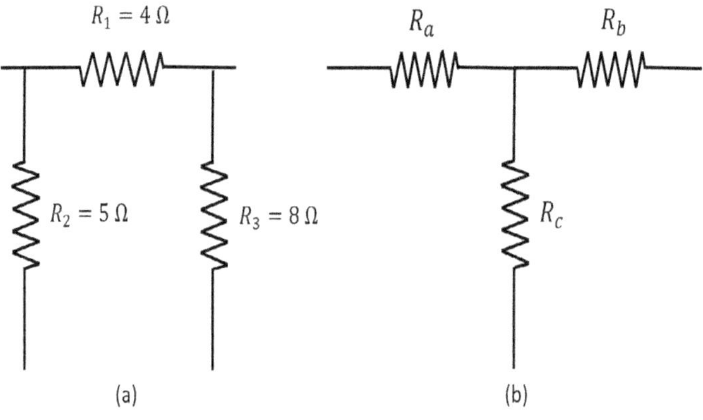

Figure 3-14. Circuit from exercise 3-16

4. ELECTRONIC COMPONENTS

Introduction

Electronics is a branch of engineering that focuses on the design, development, and application of devices and systems that use electric currents and electromagnetic fields to process, transmit, and store information. These devices and systems include components we have already seen, such as resistors, capacitors, inductors, as well as more specific ones like diodes, transistors, integrated circuits, among many others.

Electronics is used in a wide range of applications, from consumer electronics, including televisions, radios, computers, smartphones, and audio devices, to industrial, medical, military, and aerospace fields. Electronic technology has revolutionized the way we communicate, work, have fun, and explore the world.

The design of electronic systems involves understanding the fundamental principles that govern electricity, as well as the ability to use design tools and simulation software to develop and test electronic circuits.

Engineers in this field must also be capable of working in teams since designing complex electronic systems often requires collaboration with experts from different disciplines such as mechanics, computer science, and software engineering.

In the following sections, we expand the list of components we have previously seen in Chapter 2, bearing in mind that both electrical and electronic engineering are specialized branches of electricity.

Potentiometer

A potentiometer is an electronic component used to control electrical resistance in a circuit. It consists of an adjustable resistor with a movable wiper that can be moved along the resistance, increasing or decreasing its value, as shown in Figure 4-1, where you can see (a) the typical appearance and (b) the representation of a potentiometer.

This device is commonly used to adjust the volume in audio equipment such as amplifiers and sound systems. It can also be used to control the intensity of light in lamps or to regulate the speed of an electric motor.

Potentiometers are classified based on their resistance type, which can be linear or logarithmic (or antilogarithmic). Linear potentiometers have resistance that increases or decreases uniformly as the wiper moves, while logarithmic (and antilogarithmic) potentiometers have a resistance curve that increases or decreases exponentially, as shown in Figure 4-2.

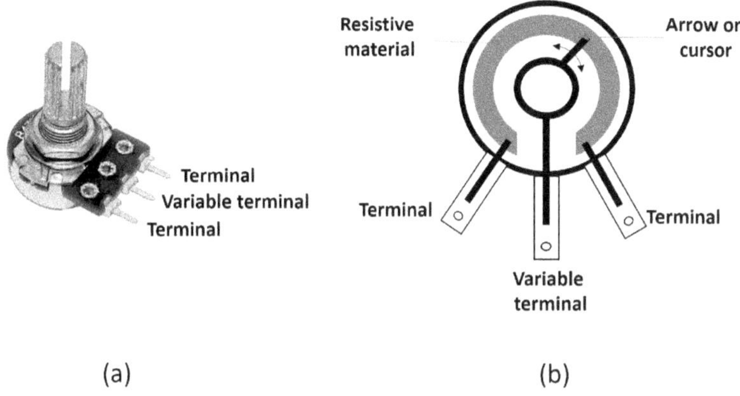

Figure 4-1 (a) Typical appearance and (b) symbol of a potentiometer

They can also vary in their power rating, which indicates the amount of electrical energy they can dissipate without being damaged. Smaller potentiometers, such as trimpots like the one shown in Figure 4-1 (a), have a low power rating and are used to adjust small amounts of current or voltage in electronic circuits. Larger potentiometers, such as panel-mounted ones, have a high power rating and are used in high-power equipment like audio amplifiers.

Diode

A diode is an electronic component that allows the flow of electric current in one direction while blocking current in the opposite direction. It is a fundamental element in electronics and is used in a wide variety of circuits.

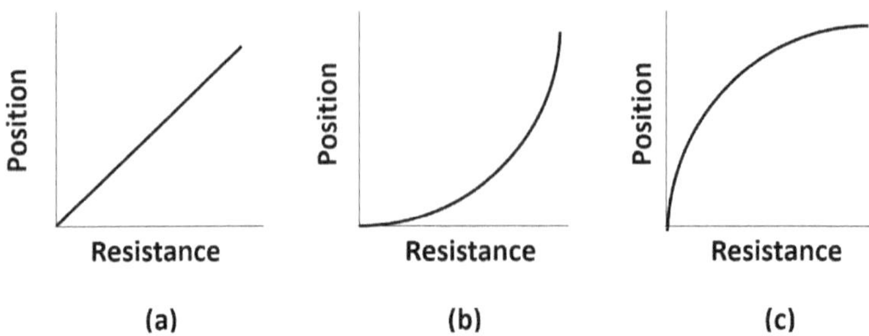

Figure 4-2 (a) Linear curve, (b) Logarithmic curve, and (c) Antilogarithmic curve of potentiometers

It is composed of a semiconductor material with unique electrical properties, making it highly useful in electronics. The semiconductor material consists of two layers, a P-type layer and an N-type layer. These layers are joined together in a region called the "PN junction." The diode has two terminals: the anode and the cathode.

When a positive voltage is applied to the P-type layer and a negative voltage is applied to the N-type layer, an electric current flows through the PN junction, and the diode allows the current to pass. On the other hand, when the polarity of the voltage is reversed, the diode blocks the electric current. It acts as a switch that opens and closes depending on the direction of the current. Refer to Figure 4-3.

Diodes have a wide variety of applications in electronics, such as current rectification, reverse polarity protection, signal detection, signal modulation, and light generation in devices like LEDs. There are also special diodes, such as Zener diodes, which have a controlled reverse voltage characteristic, and Schottky diodes, which have a lower voltage drop than conventional diodes.

The voltages of 0.3 V and 0.7 V are typical values for the

forward voltage drop of a diode. These values can vary depending on the type of diode, the semiconductor material used, and the current flowing through the diode.

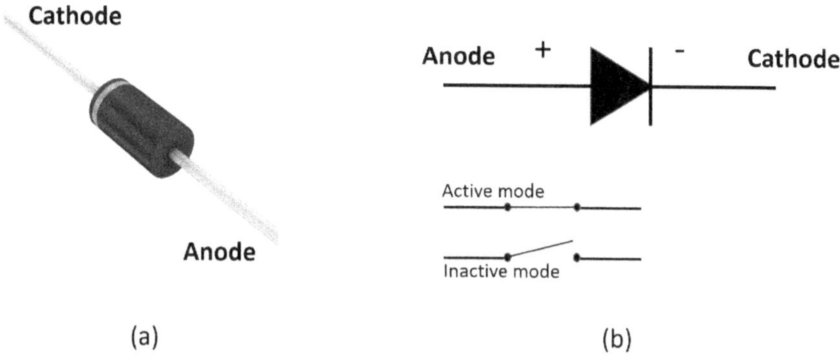

Figure 4-3 (a) Typical appearance and (b) symbol of a diode and diode behavior

In the case of the 0.3 V value, this is typical for germanium diodes and Schottky diodes, while the 0.7 V value is typical for silicon diodes and most rectifier diodes.

It is important to mention that these values are not exact and can vary within a range, so they are used as a reference for the design and calculation of electronic circuits containing diodes. Additionally, the voltage values of diodes can also vary with temperature, current, and other external factors, which should be taken into account when designing circuits containing these types of devices.

Rectifier Diode

The most typical application of a diode is as a rectifier. In this case, when a positive potential difference is applied to the anode and a negative potential difference to the cathode, the rectifier diode is forward-biased and allows the flow of current through it. However, when the polarity is reversed and a positive potential difference is applied to the cathode and a negative potential difference to the anode, the diode is reverse-biased and blocks the flow of electric current.

Half-Wave Rectifier

With an alternating current feeding the circuit, it allows the flow of current in one direction (positive) and blocks the flow in the opposite direction (negative), converting the output across the resistor into direct current, but only during each half-cycle, as shown in Figure 4-4.

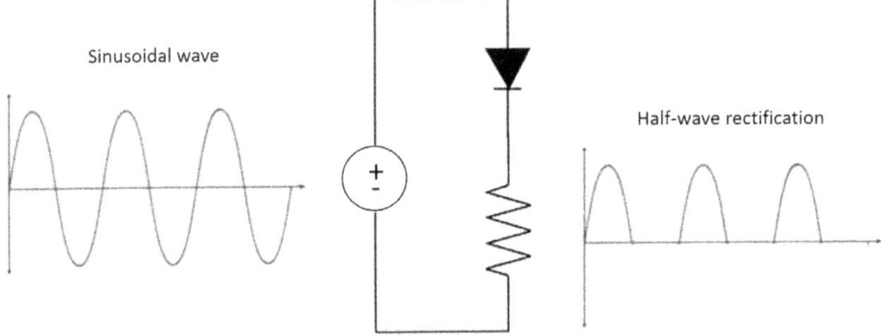

Figure 4-4. Half-wave rectifier circuit

Full-Wave Rectifier

The full-wave rectifier diode uses both cycles of the alternating current to produce smoother direct current, as shown in Figure 4-5.

Rectifier diodes are used in a wide variety of applications, from power supplies for electronic equipment to battery chargers and motor control circuits. The choice of rectifier diode type depends on the specific application and required electrical characteristics.

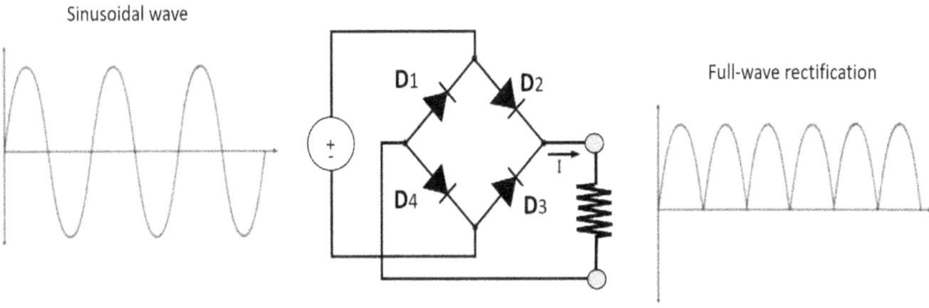

Figure 4-5. Full-wave rectifier circuit

Detector Diode

A detector diode (or small-signal diode) is a semiconductor device used to convert radiofrequency (RF) signals into direct current (DC) signals, which can be measured or amplified. Figure 4-6 shows the appearance of a detector diode or small-signal diode.

The detector diode operates in a reverse-biased configuration and relies on the nonlinear properties of the diode to rectify the input signal.

When an RF signal is applied to the reverse-biased detector diode, the current through it varies according to the amplitude of the input signal. The diode allows current to flow in only one direction, so the input signal is rectified and converted into a direct current signal.

Detector diodes are used in radio receivers, televisions, and wireless communication systems. They are also used in signal detection circuits in measurement instruments, test equipment, and process control systems.

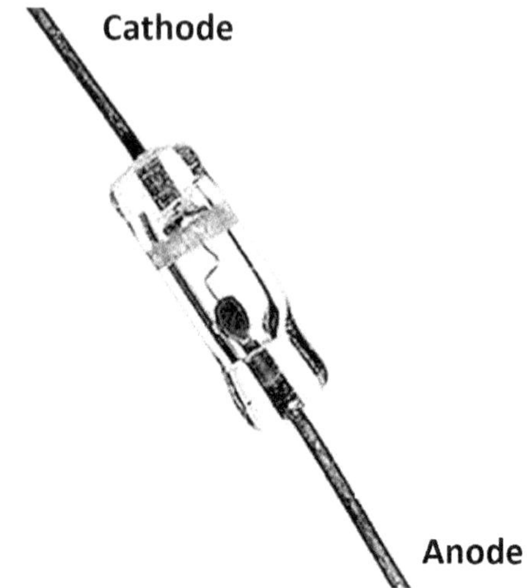

Figure 4-6. Appearance of a detector diode or low signal diode

They can be made of various materials such as germanium, silicon, and gallium arsenide. The choice of material depends on the application and the characteristics of the input signal. Germanium diodes have a high-speed response and are used in high-frequency detection applications, while silicon diodes are more common due to their lower cost and ease of fabrication.

Zener Diode

The Zener diode is a special type of diode used in voltage regulation applications and protection of electronic circuits against overvoltages. It is characterized by having a constant and predictable voltage drop, independent of the current passing through it, and its ability to operate in reverse bias at a specific voltage called the Zener voltage. Figure 4-7 shows the symbol of this component.

When reverse biased, its Zener breakdown voltage causes the diode to start conducting current in the opposite direction to the normal forward bias current. This effect allows the Zener diode to be used as a voltage regulator, maintaining a constant voltage in the circuit despite variations in current and voltage. The Zener diode is also used as protection against overvoltages in electronic circuits since once the Zener voltage is reached, current starts flowing and protects sensitive circuit components from damage.

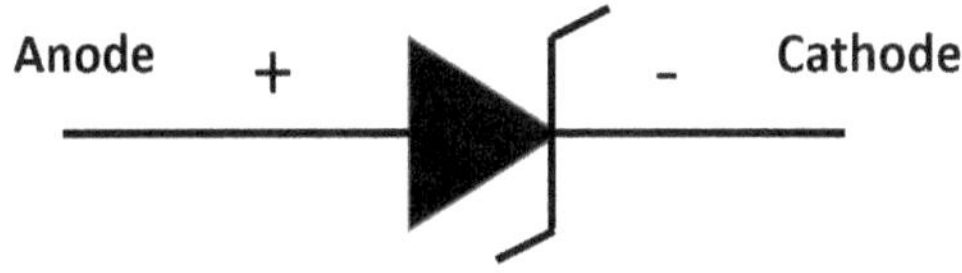

Figure 4-7. Symbol of the Zener diode

The Zener diode is used in a wide range of electronic applications, such as power supplies, voltage regulators, amplifiers, oscillators, and current limiting in protection circuits.

There are various types of Zener diodes available, each with a specific Zener voltage and different power-handling capabilities. Low-voltage Zener diodes are used in low-power voltage regulation applications, while high-power ones are used in high-power applications.

Signal Diode

The signal diode is a common type of diode used in low-power applications, such as radio frequency circuits, televisions, and other communication equipment. It is designed to have a faster recovery time and more precise switching capability than other types of diodes, and it is also characterized by its low charge storage capacity.

It is a device that has a similar construction to the standard diode, with an anode and a cathode, and is forward-biased to allow the flow of current in one direction.

The signal diode is used in rectification circuits and in modulation, demodulation, and frequency mixing applications in communication systems. It is also used in overvoltage protection applications in electronic circuits, where its fast switching capability can protect sensitive components against overvoltages.

There are various types of signal diodes available, such as Schottky diodes and PIN diodes, which have different characteristics and applications.

Schottky Diode

The Schottky diode, also known as a barrier diode, is a type of semiconductor device used to rectify alternating current (AC) signals into direct current (DC). Unlike conventional diodes, which are built with a PN junction, Schottky diodes are made of an N-type semiconductor material and a metal, forming a metal-semiconductor junction. Figure 4-8 shows the symbol of a Schottky diode.

The main advantage of Schottky diodes is their low forward voltage drop, which is due to the metal-semiconductor junction having a lower potential barrier than the conventional PN junction. This results in lower energy dissipation and higher efficiency in rectifying AC signals.

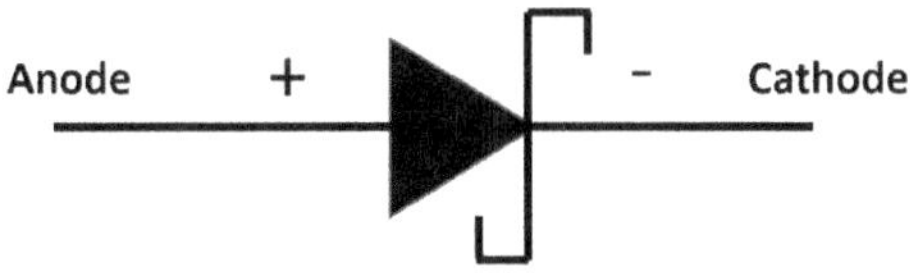

Figure 4-8. Symbol of a Schottky diode

Another advantage of these diodes is their switching speed. Due to the absence of a charge storage region in the metal-semiconductor junction, their response to input signals is faster than that of conventional PN diodes. This makes them ideal for high-frequency applications such as switching circuits and rectification of radiofrequency signals.

However, Schottky diodes also have some limitations. For example, the metal-semiconductor junction can be vulnerable to damage from high temperatures and high currents, which can degrade the device's lifespan and reliability. Additionally, their reverse voltage drop is higher than that of PN diodes, so they are not the best choice for overvoltage protection applications.

PIN Diode

The PIN diode is a type of semiconductor device commonly used in high-frequency and communication applications, such as radio frequency amplifiers, signal detectors, and modulators. Unlike conventional diodes, which have a PN junction, PIN diodes are constructed with an intrinsic region of P-type or N-type semiconductor between the doped P and N regions. Figure 4-9 shows the symbol of a PIN diode.

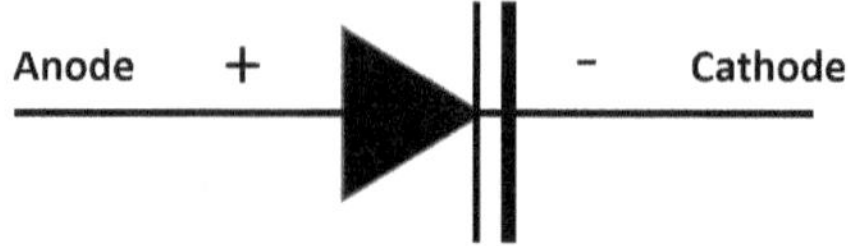

Figure 4-9. Symbol of a PIN diode

The intrinsic region in the center of the PIN diode acts as a charge region that allows the passage of charge carriers through the device more efficiently. This results in higher sensitivity and lower parasitic capacitance, enabling the PIN diode to operate at higher frequencies than common diodes.

Another advantage is its low forward impedance and high reverse impedance. Due to the large area of the intrinsic region, the forward impedance is low, resulting in lower energy loss and higher efficiency in transmitting high-frequency signals. The high reverse impedance also means that the PIN diode can withstand higher reverse voltages than conventional diodes. Additionally, PIN diodes have a more linear frequency response and higher sensitivity than common diodes. This is because the intrinsic region allows for greater capture of weak signals and greater reduction of harmonic distortion, making them ideal for high-quality signal applications, such as high-speed communication systems.

Light Emitting Diode (LED)

Light Emitting Diodes (LEDs) are semiconductor devices that

emit light when an electric current is applied to them. Figure 4-10 shows the physical appearance of LED diodes:

Figure 4-10. Physical appearance of LED diodes

They are widely used as a light source in a variety of applications, from television and mobile phone displays to traffic signage and lighting fixtures in homes and buildings. They are also used in signaling and safety applications, such as brake lights and turn indicators in automobiles. Figure 4-11 shows the symbol of an LED diode.

LEDs consist of a layer of semiconductor material sandwiched between two terminals. When an electric current is applied through the LED, electrons in the semiconductor layer recombine with electron holes (vacancies) in the same layer, releasing energy in the form of light. The color of the emitted light depends on the semiconductor material used in the active layer.

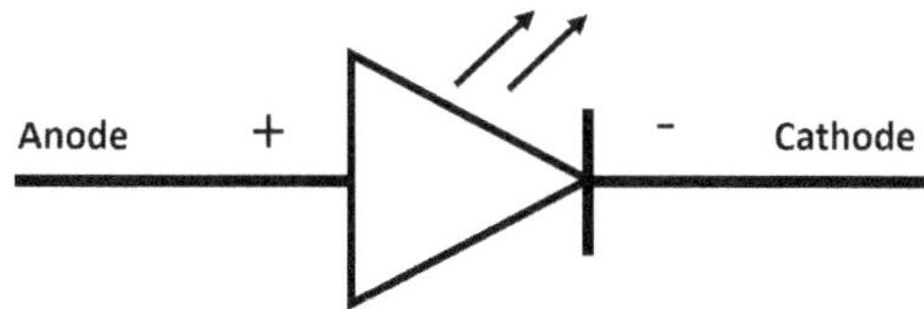

Figure 4-11. Symbol of an LED diode

In the above explanation, "electron holes" refer to regions

where atoms in the semiconductor material have a deficiency of electrons, resulting in a lack of negative charge in that region. The holes behave as positive charges, so they can combine with electrons to form an electron-hole pair, allowing current to flow through the semiconductor material.

LEDs have several advantages over other light sources such as incandescent bulbs and fluorescent lamps. Some of the advantages of LEDs include:

1. **Energy efficiency**: They are more efficient in converting electrical energy into light than incandescent bulbs and fluorescent lamps, meaning they consume less electrical energy and are more cost-effective.

2. **Durability**: They have a much longer lifespan than incandescent bulbs and fluorescent lamps, requiring less maintenance and replacement.

3. **Size and design**: They are very small and can be manufactured in a variety of shapes and sizes, making them suitable for use in a wide range of applications.

4. **Light quality**: They emit brighter and more uniform light than incandescent bulbs and fluorescent lamps, and can be adjusted to produce different tones and colors of light.

Infrared Light Emitting Diode

An infrared light-emitting diode, or infrared LED, is an electronic device that emits light in the infrared spectrum range, which has longer wavelengths than visible light. Figure 4-12 shows the symbol of an infrared LED diode, note that it is the same as that of a conventional LED diode:

Infrared LEDs are based on the same principle as conventional LEDs, but instead of emitting visible light, they emit infrared light. These diodes are made of semiconductor materials, such as gallium arsenide and germanium-doped silicon, which have an appropriate energy band to produce photon emission in the infrared range.

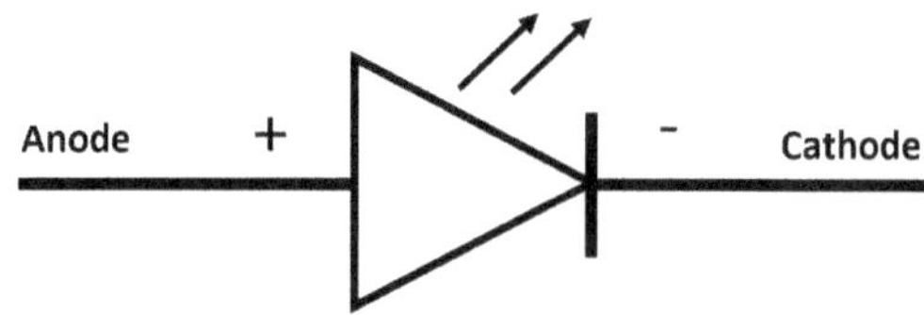

Figure 4-12. Symbol of an infrared LED diode

The process of light emission in an infrared LED is similar to that of a conventional LED. When a voltage is applied across the diode, electrons move from the N-type region to the P-type region of the diode, where they combine with the electron holes (as mentioned before). This recombination of electrons and holes releases energy in the form of infrared light photons.

Infrared LEDs are used in many remote control applications, such as television systems and DVD players, to transmit infrared signals to a receiver located in the device being controlled. They are also used in motion sensors and security systems, where the presence of people or objects is detected using reflected infrared light. Additionally, they are used in optical communication equipment and networks to transmit information signals through fiber optic cables.

Photodiodes

A photodiode is a semiconductor device that converts light into an electric current. It is a type of diode used in light detection applications, such as digital cameras, motion sensors, security systems, and telecommunications equipment. Figure 4-13 shows the symbol of a photodiode:

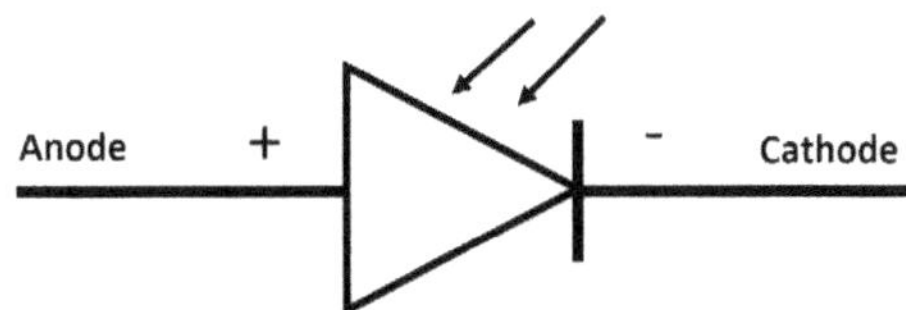

Figure 4-13. Symbol of a photodiode

The operation of a photodiode is similar to that of a conventional diode, but instead of allowing current to pass in only one direction, photodiodes generate an electric current when light shines on their surface. The energy of the light photons is transferred to the electrons in the semiconductor layer, creating electron-hole pairs in the PN junction region of the photodiode. These pairs are separated by the internal electric field of the device, resulting in an electric current flowing through it.

Photodiodes are manufactured using semiconductor materials such as silicon and germanium and can be of different types, such as avalanche photodiodes and surface passivation photodiodes. Avalanche photodiodes are more sensitive to light and have internal gain, meaning that a small amount of light can generate a large current. Surface passivation photodiodes, on the other hand, have higher efficiency and lower noise, making them ideal for high-sensitivity applications.

Photodiodes are used in a wide range of applications, from

light detection in digital cameras to measuring light intensity in industrial automation systems. They are also used in tele-communications equipment for receiving optical signals in fiber optic cables.

Switch

A switch is an electronic component used to open or close an electrical circuit, thereby controlling the flow of current that passes through or stops through part or all of the circuit, making it work or stop working.

There are various types of switches, classified based on their design and function. Some of the most common ones are shown in Figure 4-14:

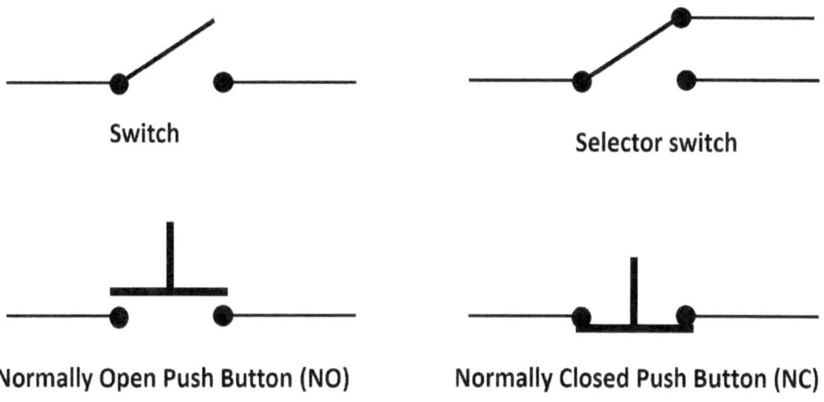

Figure 4-14. Types of switches

Switches are used in a wide variety of applications, from small electronic devices to large-scale electrical systems. Some of the most common applications include lighting control, motor activation, and opening/closing of electric doors and windows.

Fuse

Fuses are protective devices used to interrupt the flow of electric current in a circuit when the current exceeds a safe level. These devices consist of a metal conductor (filament) placed in series with the circuit to be protected. When the current in the circuit exceeds a safe level, the conductor heats up and melts, interrupting the current flow and protecting the circuit from possible damage. Figure 4-15 shows the appearance of an electronic fuse.

Fuses are used in a wide range of applications, from electronic devices to large-scale electrical systems. For example, fuses can be found in household electrical outlets to protect appliances from potential electrical overload. They are also used in motor protection systems and lighting circuits in homes and automobiles.

Figure 4-15. Fuse

There are different types of fuses that can be tailored to the specific needs of each application. They can be classified based on their current capacity, response speed, nominal voltage, casing type, among other factors. Additionally, some fuse systems may be replaceable, while others are designed to break irreversibly, requiring the entire system to be replaced with a new one.

Transistor

The term "transistor" refers to a semiconductor device used to control the flow of electric current in a circuit. There are various types of transistors, including the bipolar transistor and the Field Effect Transistor (FET). These two types of transistors can be seen in Figure 4-16.

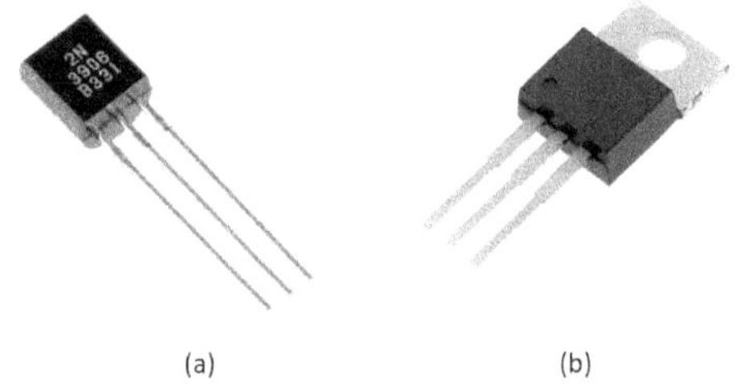

Figura 4-16. Transistors (a) Bipolar, (b) FET

Bipolar Transistor

The bipolar transistor is commonly used in electronics as a signal amplifier, switch, and a key element in integrated circuit construction. It consists of three regions of P-type and N-type semiconductors, known as the emitter (E), base (B), and collector (C), which are connected to form two PN junctions. Consequently, there are NPN and PNP type transistors. The symbols for these two types of transistors are shown in Figure 4-17.

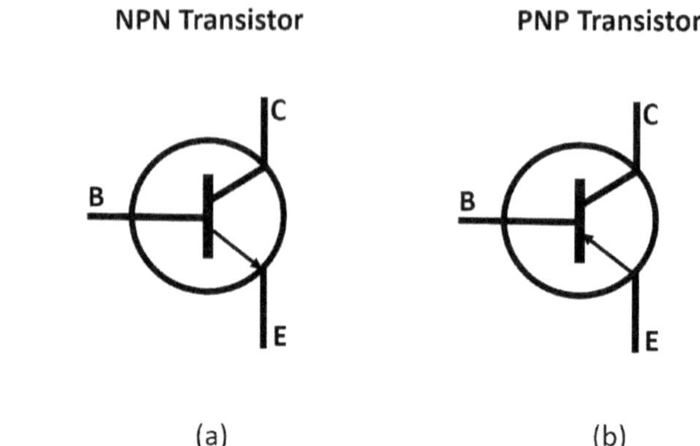

Figure 4-17. Symbol of a transistor (a) NPN type and (b) PNP type

The principle of operation for the bipolar transistor is based on the ability of the base to control the flow of current from the emitter to the collector. When a current is applied to the base, a current flow occurs between the emitter and the collector. The amount of current flowing is determined by the current applied to the base.

When a small current is applied to the base, the transistor can amplify the current flowing through the collector. This makes it ideal for signal amplification applications, such as in audio amplifiers and radio frequency circuits.

The transistor is also used as a switch. In this case, a current is applied to the base to turn on the transistor and allow current to flow through the collector. By removing the current from the base, the device turns off, ceasing the flow of current through the collector. This makes it suitable for switching applications, such as motor and light control circuits..

Transistors are also used in integrated circuit construction. These integrated circuits are composed of many electronic elements, including transistors, which have been miniaturized and placed on a silicon chip. This allows complex electronic circuits to be built in a very small space, making integrated circuits ideal for use in portable electronic devices and computers.

Bipolar transistors are older than Field Effect Transistors (FETs), but they remain important in many electronic circuits due to their high-speed amplification and switching properties.

Field Effect Transistor (FET)

The Field Effect Transistor (FET) is a semiconductor component commonly used in electronics for signal amplification, circuit switching, and as a key element in integrated circuit construction. The FET operates by controlling the current through a semiconductor conducting channel via an electric field generated by an insulated gate.

There are two main types of FETs: the N-channel FET and the P-channel FET. Both types have a similar structure consisting of a source (S), a drain (D), and a gate (G). The difference lies in the type of semiconductor material used in the conducting channel; the N-channel FET has an N-type semiconductor region, while the P-channel FET has a P-type semiconductor region. The symbols for these two FETs are shown in Figure 4-18, (a) for the N-channel FET, and (b) for the P-channel FET.

CWhen a voltage is applied to the gate, an electric field is generated that modulates the conductivity of the conducting channel. If the gate voltage is positive compared to the source (in the case of the N-channel FET), negative charges are attracted to the conducting channel, increasing its conductivity and allowing more current to flow between the drain and the source. If the gate voltage is negative, negative charges are repelled from the conducting channel, reducing its conductivity and decreasing the amount of current flowing between the drain and the source.

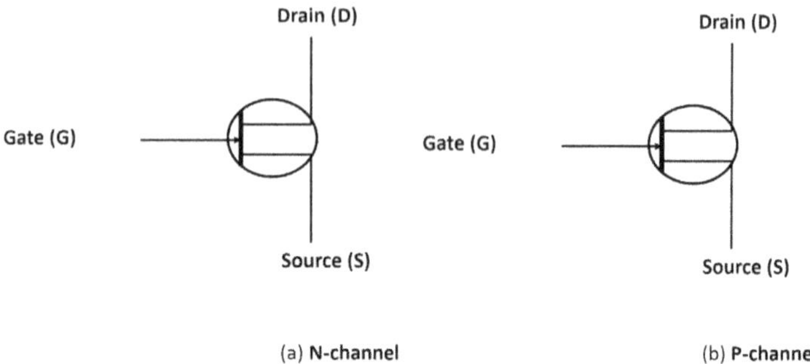

Figure 4-18. Symbol of a FET transistor (a) N-channel and (b) P-channel

FETs offer several advantages over other electronic devices, such as bipolar transistors, due to their very high input resistance. This means that a very small input current is sufficient to control the current through the device, making FETs more power-efficient. Additionally, FETs have a high input impedance, making them sensitive to very small signals and suitable for low-level signal amplification applications. Lastly, FETs have low harmonic distortion, meaning they introduce very little distortion to the input signal, making them suitable for high-fidelity applications.

Differences between Bipolar Transistor and FET

The main difference between a bipolar transistor and a Field Effect Transistor lies in the mechanism by which they control the current flow. In a bipolar transistor, current is controlled by the current applied to the base of the transistor. This makes bipolar transistors relatively fast but less energy-efficient. In contrast, in a FET, current is controlled by an electric field generated by a voltage applied to the control terminal, the gate. Because of this, FETs have a very high input impedance, requiring very little input current to function, making them more energy-efficient.

Another significant difference is that bipolar transistors can have very high current gain, while FETs have much lower current gain. This makes bipolar transistors more suitable for applications that require high gain, such as amplifiers, whereas FETs are more suitable for applications that require high input impedance, such as instrumentation amplifiers and sensors.

Phototransistor

It is a type of transistor that is activated by light. It is designed to detect light and convert it into an electrical signal. It consists of a conventional transistor with a transparent window in the collector region to allow light to reach the collector-base junction. In Figure 4-19, the symbols for the most common use of NPN and PNP phototransistors can be seen.

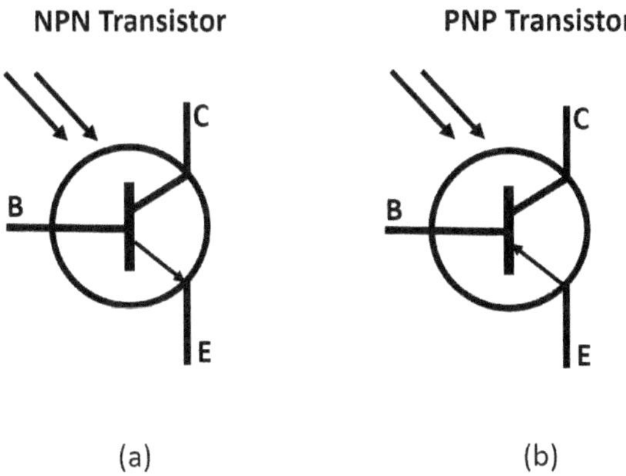

Figure 4-19. Symbol of a phototransistor (a) NPN type and (b) PNP type.

When light falls on the collector window, the amount of current flowing through the transistor increases, resulting in an electrical signal. The phototransistor essentially acts as a switch that is activated by the presence of light on the collector window.

It is commonly used in light detection applications, such as environmental sensors and infrared remote controls. It is also used in communication applications, such as transmitting television signals through optical fibers.

One of its advantages is its high sensitivity. As a result, it can detect very low levels of light and convert them into useful electrical signals. Additionally, it is faster than other similar devices, such as photodiodes, making it suitable for high-speed applications.

Sensor

Sensors are devices used to detect, measure, and convert physical quantities into electrical signals. These quantities can be of different types, such as temperature, pressure, light, humidity, motion, among others.

They are used in a wide variety of applications, from the automotive and aerospace industries to medicine, agriculture, security, and smart homes. Some examples of common sensors include:

1. **Temperature sensor**: used to measure the temperature of an object or environment and convert it into an electrical signal.

2. **Pressure sensor**: used to measure pressure in a system or environment and convert it into an electrical signal.

3. **Light sensor**: used to measure the amount of light in an environment or to detect the presence or absence of light.

4. **Motion sensor**: used to detect the movement of people, animals, or objects in an environment.

5. **Humidity sensor**: used to measure the amount of moisture in an environment or object.

6. **Gas sensor**: used to detect the presence of hazardous or toxic gases in the air.

Sensors can be analog or digital, and their operation depends on the type of quantity they measure. Some can be wireless, meaning they can send data to other devices without the need for physical wiring, for analysis and processing.

Microcontroller

Microcontrollers are electronic devices that integrate a microprocessor, memory, and input/output peripherals into a single chip. They are widely used in the electronics industry for the control and automation of systems, as well as in consumer electronics such as appliances, toys, and mobile devices.

The microprocessor is the main component of a microcontroller, responsible for processing the instructions of the program stored in memory. This memory can be of two types: read-only memory (ROM) or read/write memory (RAM). ROM is used to store the program and non-changing data, while RAM is used to store changing data. In Figure 4-20, (a) shows the typical physical appearance of a microprocessor, and (b) the internal architecture of a microcontroller: though it may seem simple, the components that interact are the basis of all the digital technology we know and use today.

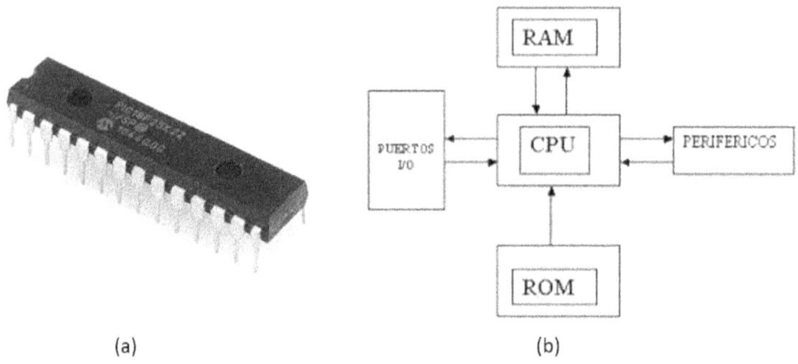

Figure 4-20 (a) Physical appearance and (b) architecture of a microcontroller

The input/output peripherals are the components that allow the microcontroller to communicate with the outside world. Among them are input/output ports, which allow the connection of external sensors and actuators, and analog-to-digital converters (ADC) and digital-to-analog converters (DAC), which allow the conversion of analog signals to digital and vice versa.

Microcontrollers are programmed using high-level programming languages such as C and C++, or specific languages such as assembler. Programs can be loaded into the microcontroller using programming interfaces that connect the microcontroller to a computer.

Signal Converters

A signal converter is a device that converts a signal from one format to another. The purpose of such an element is to enable interoperability between components or systems that use different signal formats.

Some of the most common signal converters include:

1. **Analog-to-digital converters (ADC):** convert an analog signal into a digital signal, which can be processed by devices such as computers.

2. **Digital-to-analog converters (DAC):** convert a digital signal into an analog signal, which can be processed by other devices, such as speakers.

3. **Frequency converters:** convert a signal from one frequency to another. Such a component can be used to change a high-frequency radio signal to a low-frequency audio signal.

4. **Voltage converters**: convert a signal voltage from one level to another. For example, it can be used to change an audio signal from a line level to a microphone signal level.

5. **Protocol converters:** convert a signal from one protocol to another. For example, it can be used to convert an Ethernet signal to an optical fiber signal.

In this case, we will focus on the first two.

Analog-to-Digital Converter (ADC)

As we mentioned before, an analog-to-digital converter (ADC) is an electronic device that converts an analog signal, such as the one obtained from a temperature, pressure, or light sensor (analog), into a digital signal to be processed by a microcontroller or a computer (digital). The analog signal is continuous in time and amplitude, while the digital signal is discrete in time and amplitude. In Figure 4-21, we see an analog sinusoidal signal that, when passed through an ADC, generates another signal in digital form.

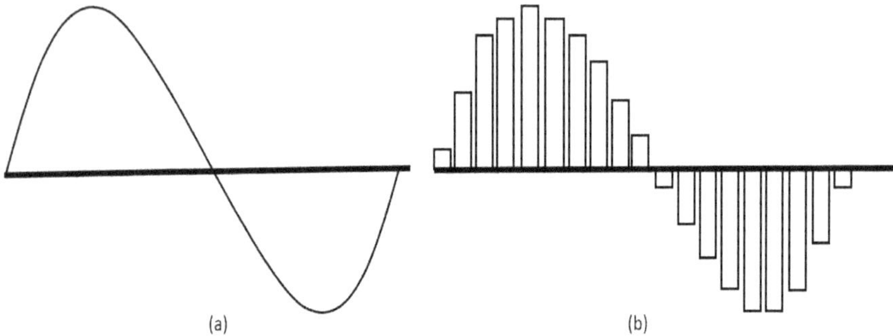

Figure 4-21 Signal conversion (a) analog to (b) digital

The conversion process is carried out in two stages: sampling and quantization.

1. During the sampling stage, the ADC takes samples of the analog signal at regular time intervals. The faster the samples are taken, the higher the temporal resolution of the resulting digital signal.

2. During the quantization stage, the ADC assigns a digital value to each sample of the analog signal. The higher the resolution of the ADC, the more accurate the digital values assigned to each sample.

The resolution of an ADC is measured in bits and determines the number of discrete values the resulting digital signal can take. For example, an 8-bit ADC can take up to 256 discrete values, while a 16-bit ADC can take up to 65,536 discrete values.

There are different types of ADCs, such as successive approximation ADCs, ramp ADCs, flash ADCs, among others. Each type of ADC has its advantages and disadvantages in terms of speed, resolution, and cost.

Digital-to-Analog Converter (DAC)

A digital-to-analog converter (DAC) is an electronic device that converts a digital signal into an analog signal. The digital signal consists of a series of discrete values, while the analog signal is continuous in both time and amplitude. As shown in Figure 4-22, from a (a) digital signal, composed of a large number of discrete values, a (b) analog signal is obtained.

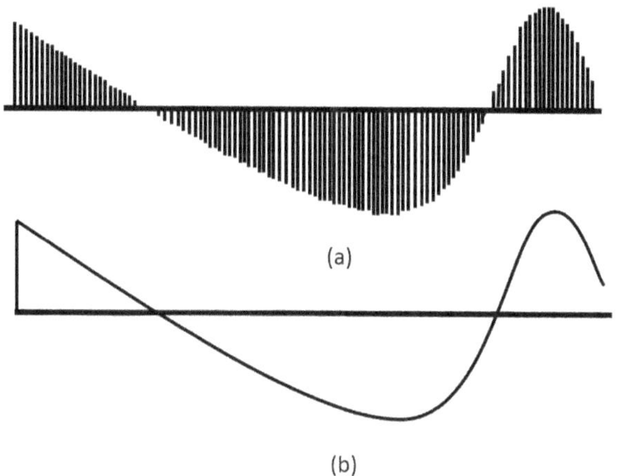

Figure 4-22.Signal conversion (a) digital to (b) analog

A DAC is commonly used in applications such as audio and video reproduction, motor control, and automation system control. In these applications, the digital signal is processed in a microcontroller or computer and converted into an analog signal used to control the system.

In the digital-to-analog conversion process, there are two stages: sampling and reconstruction.

1. During the sampling stage, the DAC takes a series of digital values and converts them into a continuous analog signal. The more frequent the digital values, the higher the temporal resolution of the resulting analog signal.

2. During the reconstruction stage, the DAC smoothens the resulting analog signal to eliminate the effects of discontinuities in the original digital signal.

There are different types of DACs, such as weighted resistor DACs, R-2R ladder network DACs, and sigma-delta DACs. Each type of DAC has its advantages and disadvantages in terms of speed, resolution, and cost.

Integrated Circuit

An integrated circuit (IC) is an electronic component composed of various electronic devices such as transistors, diodes, and resistors, interconnected on a single silicon chip (as shown in Figure 4-20 (a)). Integrated circuits are used to perform a variety of functions in modern electronic systems, from signal processing to data communication and motor control.

Integrated circuits are classified into two main categories: analog and digital. The former is used to process analog signals, such as audio and radio signals, while the latter is used to process digital signals, such as those handled by a computer.

These devices can also be classified based on their complexity. Simple integrated circuits, such as operational amplifiers (which we will see next) and comparators, consist of a few transistors and other electronic components. More complex integrated circuits, such as microprocessors and device controllers, can contain millions of transistors and other electronic components.

Integrated circuits are manufactured using semiconductor processing techniques that involve depositing thin layers of semiconductor materials, such as silicon, onto crystal substrates. The circuit patterns are created using lithography, which uses light to engrave patterns on the semiconductor material layers. Integrated circuits can be encapsulated in various forms, from flat packages to plastic packages with pins.

The IC - Operational Amplifier

An operational amplifier (also known as op-amp) is an electronic device used to amplify an electrical signal. It is an integrated circuit consisting of several transistors and other electronic components interconnected in such a way that it can be used to perform a wide variety of signal processing functions.

The operational amplifier is commonly used in signal processing applications, such as audio and control in industrial automation systems. Also, in feedback control circuits, such as voltage and current control, and signal conversion circuits, from analog to digitals.

The operational amplifier has two inputs, an inverting (-) and a non-inverting (+) input, and one output. Amplification is achieved by comparing the input voltages and applying a specific gain. The gain can be adjusted by selecting the appropriate resistance in the feedback circuit. In Figure 4-23, the representation of an operational amplifier is shown.

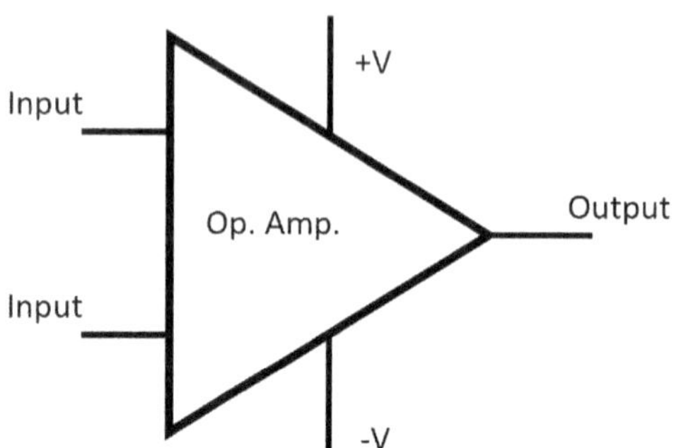

Figure 4-23. Symbol of an Operational Amplifier

In addition to signal amplification, the operational amplifier is also used to perform mathematical functions such as addition, subtraction, multiplication, integration, and differentiation. Operational amplifiers can also be used in oscillator circuits, filters, and comparison circuits.

The IC – 555

It is an integrated circuit frequently used in electronics as a timer and oscillator. It was originally designed in 1971 by Swiss engineer Hans R. Camenzind and has become one of the most popular and versatile integrated circuits in history.

It has eight pins and can be used in different ways to create various types of circuits. In Figure 4-24, the 555 IC chip and the function of each pin are shown.

In timer mode, the 555 is used to generate an output pulse after a certain time when an input pulse is applied. In oscillator mode, the 555 is used to generate a periodic output signal with a specific frequency.

The 555 IC can be used in a wide variety of applications, such as alarm systems, automatic control systems, switched power supplies, square wave generators, lighting systems, among others. Additionally, it can be used in combination with other components to create more complex circuits.

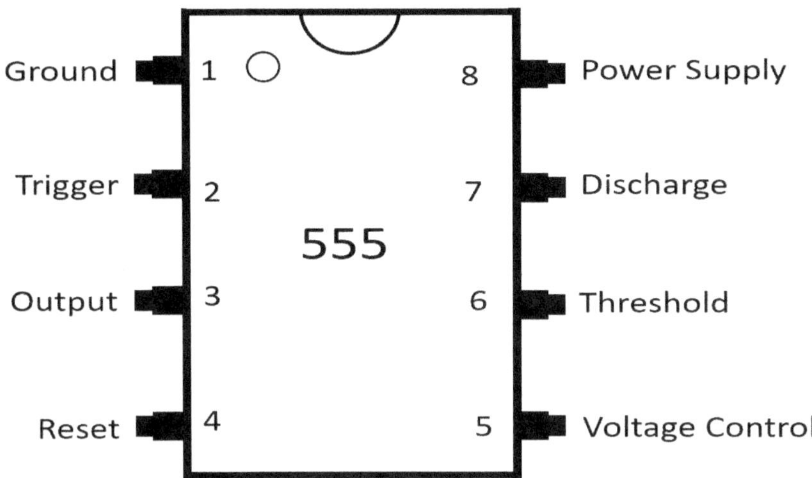

Figure 4-24. Pin functions in the IC 555

Review Questionnaire

Answer the question:

1. What is electronics?

2. What is the function of a potentiometer in an electronic circuit?

3. What types of diodes exist?

4. What is the function of a transistor in an electronic circuit?

5. What is an integrated circuit?

Select the correct option:

6. What type of diode is used to convert alternating current into direct current?
   a. Rectifier diode
   b. Zener diode
   c. Signal diode
   d. Schottky diode

7. What is the name of the rectifier that uses only half of the input wave?
   a. Full-wave rectifier
   b. Half-wave rectifier
   c. Zener rectifier
   d. Rectifier diode

8. What is a diode detector used for?
   a. To rectify the input signal
   b. To detect changes in current
   c. To detect changes in voltage
   d. To attenuate the input signal

9. What type of diode is used to maintain a constant voltage in a circuit?
   a. Rectifier diode
   b. Zener diode
   c. Signal diode
   d. Schottky diode

10. What type of diode is used for high-frequency applications?
    a. Rectifier diode
    b. Zener diode
    c. Signal diode
    d. Schottky diode

11. What type of diode emits visible light when forward-biased?
    a. Light Emitting Diode (LED)
    b. Infrared Light Emitting Diode
    c. Photodiodes
    d. Switch

12. What type of diode is used to detect light?
    a. Light Emitting Diode (LED)
    b. Infrared Light Emitting Diode
    c. Photodiodes
    d. Switch

Answer if the statement is true or false:

13. Microcontrollers are devices that can perform programmed tasks and control the operation of other devices and systems.
    a. True
    b. Falso?

14. Signal converters are devices that convert an analog electrical signal into a digital signal and vice versa.
    c. True
    d. Falso?

15. An ADC converts an analog signal into a digital signal, while a DAC converts a digital signal into an analog signal.
    a. True
    b. Falso?

16. An integrated circuit is an electronic device that contains various electronic components, such as resistors, capacitors, transistors, and diodes, on a single silicon chip.
    a. True
    b. Falso?

17. An operational amplifier is an integrated circuit used to amplify electrical signals and can be configured to perform different functions such as adder, subtractor, integrator, differentiator, etc.
    a. True
    b. Falso?

18. The 555 integrated circuit is a timer that can be used in a variety of applications, such as pulse generator, oscillator, timer, among others.
    a. True
    b. Falso?

Complete the statement with the correct word or expression:

19. A __________ is an electronic component used to adjust electrical resistance in a circuit.

20. A __________ is a device used to open or close an electrical circuit.

21. A __________ is a safety component used to protect electrical circuits from current overloads.

22. A __________ is an electronic component used to amplify or control the flow of current in a circuit.

23. A __________ is a type of transistor used to amplify electrical signals, and is composed of three doped semiconductor regions.

24. A __________ is a type of transistor used to control the flow of electric current through the electric field generated by an applied voltage.

25. A __________ is a device used to measure a physical quantity and convert it into an electrical signal, which can be processed by other electronic devices.

5. DIGITAL ELECTRONICS

Introduction

Digital electronics is a branch of electronics that deals with the processing, storage, and transmission of information in the form of digital signals. It differs from analog electronics, which works with continuous signals, while digital electronics works with discrete signals, as shown in Figure 5-1.

In digital electronics, information is represented in the form of bits, which can have two possible values: 0 or 1. These bits are combined to form numbers, letters, and other types of information, and are processed using digital circuits.

Digital circuits are based on logic gates, which are electronic devices that perform basic logical operations, such as negation, conjunction, and disjunction. At the same time, these logic gates are combined to form more complex circuits, such as registers, counters, decoders, and multiplexers.

One of the main benefits of digital electronics is that it allows the manipulation and processing of large amounts of information quickly and accurately. Additionally, it is easy to integrate into complex systems and can be programmed to perform a wide variety of tasks.

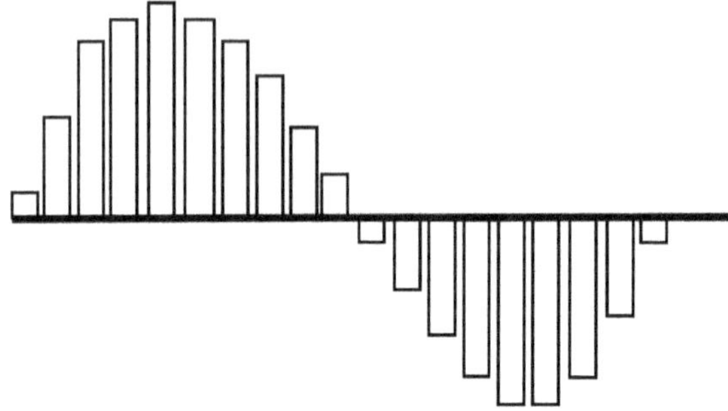

Figure 5-1. Sinusoidal wave composed of discrete signals (pulses)

Digital electronics is applied in a wide range of fields, such as computer science, telecommunications, industrial process control, robotics, consumer electronics, building automation, and many others.

Number Systems

Number systems are used to represent numerical quantities. The most common number systems are decimal, binary, octal, and hexadecimal.

Decimal System

The decimal system is based on the number 10, meaning it uses 10 different digits (0 to 9) to represent all numerical quantities. Each digit in a decimal position has a different value, depending on that position. For example, the number 1234 in the decimal system is expressed as follows:

$$(1x10^3) + (2x10^2) + (3x10^1) + (4x10^0) =$$

$$(1x1000) + (2x100) + (3x10) + (4x1) =$$

$$1000 + 200 + 30 + 4 =$$

$$1234$$

Binary System

The binary system is based on the number 2 and, as already indicated, uses only two digits (0 and 1) to represent all numerical quantities. In the binary system, each digit in a binary position has a different value, depending on that position. For example, the number 1101 in the binary system means in the decimal system:

$$(1x2^3) + (1x2^2) + (0x2^1) + (1x1^0) =$$

$$(1x8) + (1x4) + (0x2) + (1x1) =$$

$$8 + 4 + 0 + 1 =$$

$$13 \text{ (en el sistema decimal).}$$

Octal System

The octal system is based on the number 8 and uses eight different digits (0 to 7) to represent all numerical quantities. Each digit in an octal position has a different value, depending on the position. For example, the number 375 in the octal system is expressed in the decimal system as follows:

$$(3x8^2) + (7x8^1) + (5x8^0) =$$

$(3x64) + (7x8) + (5x1) =$

$192 + 56 + 5$

253 (en el sistema decimal)

Hexadecimal System

The hexadecimal system is based on the number 16 and uses 16 different digits (0 to 9 and A to F, where: A=10, B=11, C=12, D=13, E=14, F=15) to represent all numerical quantities. Each digit in a hexadecimal position has a different value, depending on the position. For example, the number A5 in the hexadecimal system means in the decimal system:

Considering that A=10,

$(10x16^1) + (5x16^0) =$

$10x16 + 5x1 =$

$160 + 5 =$

165 (in the decimal system)

Each number system has its own advantages and disadvantages and is used in different applications. The decimal system is the most commonly used in everyday life, while the binary system is widely used in electronics and computer science. The octal and hexadecimal systems are used in computer programming and in the representation of memory addresses.

Boolean Algebra

Boolean algebra is an area of mathematics and electronics that deals with the study and manipulation of algebraic expressions containing variables and logical operations. These logical variables can have only two possible values: true (1) or false (0).

The most common logical operations are negation (NOT), conjunction (AND), and disjunction (OR).

Negation is used to invert the logical value of a variable; for example, NOT A means that variable A changes from 1 to 0 or from 0 to 1.

Conjunction is used to combine two or more logical variables and is true only if all the variables are true; for example, A AND B is true only if both A and B are true.

Disjunction is used to combine two or more logical variables and is true if at least one of the variables is true; for example, A OR B is true if A is true, B is true, or both are true.

Boolean algebra is used in digital electronics to design logic circuits and control systems. Logic circuits are built using logic gates, which implement basic logical operations (NOT, AND, OR). These logic gates are combined to form more complex circuits, which can perform more complex operations.

Boolean algebra is also used in computer programming, especially in control systems and computer theory. Computer programs often use logical operations to make decisions and control the flow of data:

Table 5-1. Information system for operations AND, OR and NOT

P	Q	P AND Q	P OR Q	NOT P
0	0	0	0	1
0	1	0	1	1

| 1 | O | O | 1 | O |
| 1 | 1 | 1 | 1 | O |

This table represents the logical values (true or false) of the AND, OR, and NOT operations between two variables P and Q.

In the P AND Q column, the result of the logical "AND" operation between P and Q is shown. The AND operation returns true (1) only if both variables are true.

In the P OR Q column, the result of the logical "OR" operation between P and Q is shown. The OR operation returns true (1) if at least one of the variables is true.

In the NOT P column, the result of the logical "NOT" operation on variable P is shown. The NOT operation returns true (1) if the variable is false and vice versa.

Once the basic logical operations are understood, you can begin to work on the simplification of Boolean expressions, for which laws and rules such as the law of identity, the law of complement, and De Morgan's law can be used. However, despite their importance, we will not see them in this book, as they are not part of its scope.

Isomorphisms

Isomorphisms are an important tool in mathematical theory for studying algebraic structures. In simple terms, an isomorphism is a bijective correspondence between two algebraic structures that preserves the fundamental properties of these structures.

Formally, let A and B be two algebraic structures of the same type; an isomorphism between A and B is a bijective function f: A → B that satisfies the following conditions:

f(a * b) = f(a) * f(b) for all a, b in A (where * is the binary operation of A).

f(1_A) = 1_B (where 1_A and 1_B are the neutral elements of A and B, respectively).

The first condition guarantees that f preserves the algebraic structure of A in B, i.e., the binary operation in A corresponds to the binary operation in B. The second condition ensures that f preserves the identity of the algebraic structure, i.e., the neutral element in A corresponds to the neutral element in B.

A simple example of an isomorphism is the bijective correspondence between the set of integers and the set of even numbers, defined by f(x) = 2x. This function satisfies the two conditions of an isomorphism, as f(x + y) = 2(x + y) = 2x + 2y = f(x) + f(y) for all x, y in integers, and f(0) = 2*0 = 0, which is the neutral element of even numbers.

Isomorphisms are important because they allow us to identify common properties in apparently distinct algebraic structures. For example, two groups may seem different at first glance, but if there is an isomorphism between them, then we know that they share fundamental properties such as the existence of a neutral element and the invertibility of their elements.

Isomorphisms in Digital Electronics

In digital electronics, isomorphisms are used to identify equivalent logic circuits that perform the same function, even if they may be designed differently. In other words, two logic circuits can be isomorphic if they have the same algebraic structure and perform the same boolean operation.

For example, consider two logic circuits that implement the same logic function but have different designs and use different logic gates. We can use Boolean algebra to show that these circuits are equivalent and, therefore, isomorphic.

To do this, we can construct a truth table that shows the input and output values for both circuits and then simplify the resulting Boolean expression using the laws of Boolean algebra. If both Boolean expressions are identical, then we can conclude that the circuits are equivalent and, therefore, isomorphic.

Isomorphisms are also important in optimizing logic circuits, as they allow us to identify simpler and more efficient equivalent circuits that perform the same boolean function. This can be especially useful in the design of high-speed or low-power circuits, where circuit optimization is essential.

Propositional Logic and Digital Electronics

The analogy between propositional logic and digital electronics is a fundamental concept in computer science and modern technology. This analogy is based on the idea that both propositional logic and digital electronics operate on a set of two binary states or values, allowing their application in areas such as logic circuit design, programming, and software engineering.

In propositional logic, a proposition can have only two truth values: true or false. These truth values are represented with symbols like "1" for true and "0" for false. Similarly, in digital electronics, a system or device can be in one of two binary states: "on" or "off," "open" or "closed," "high" or "low," "true" or "false," and are represented with the same numbers "1" and "o.".

This analogy allows the application of propositional logic in digital electronics, as logic circuits and digital systems are built from components that have binary states, allowing the design and construction of logic circuits and digital systems.

For example, logical operations such as AND, OR, and NOT can be used to combine binary inputs and produce binary outputs in a logic circuit. Similarly, in programming and software engineering, Boolean expressions are used to control the flow of execution in a program and make decisions based on true or false logical values. The table 5-2 illustrates this concept:

Table 5-2. Information System "Pilot State vs Information Behavior"

Truth state	Pilot state	Information behavior
1	1	1
1	O	1
O	1	O
O	O	O

The table shows the four possible states that an information system can be in and how they relate to the state of a pilot and the behavior of the information itself.

When the truth state is 1, the information passes, regardless of whether the pilot is on or off. On the other hand, when the truth state is 0, the information does not pass, regardless of the state of the pilot.

This table illustrates how the binary states of an information system relate to the binary states of a pilot and how Boolean logic can be used to describe the behavior of information systems.

Logic Circuits

Logic circuits are electronic systems that operate based on logical and mathematical principles and are used in the construction of digital devices such as computers, smartphones, gaming consoles, among others.

In Figure 5-2, we see a circuit composed of a power source feeding a lamp, but with two switches in series, which can be, any of the two, or both at the same time, in two possible states: "on" and "off."

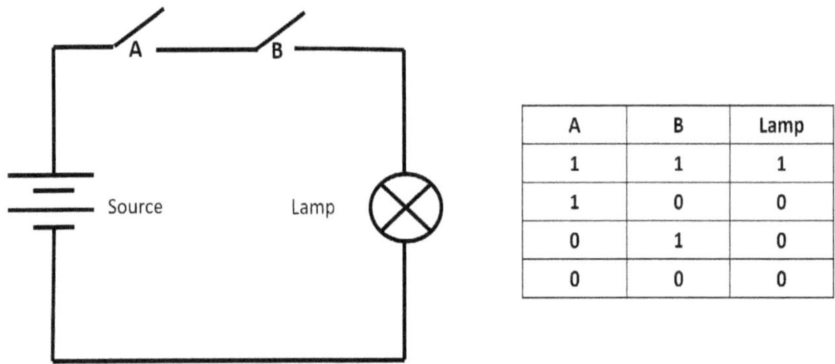

A	B	Lamp
1	1	1
1	0	0
0	1	0
0	0	0

Figure 5-2. Circuit with double switch and truth table

Logic circuits are based on two types of electronic devices: logic gates and flip-flops, and can be constructed using different technologies, such as bipolar transistors, field-effect transistors (FETs), integrated circuits, and other electronic components.

Logic gates are devices that perform simple logical operations, such as negation, conjunction, and disjunction. Flip-flops are devices used to store a bit of information and build sequential circuits.

There are different types of logic circuits, classified according to their function and complexity. Some of the most common ones include combinational circuits, which perform real-time logical operations and do not store information; sequential circuits, which store information and are used in applications like counters and registers; and arithmetic circuits, which perform mathematical operations such as addition, subtraction, multiplication, and division.

Logic circuits are essential in digital electronics, as they enable the construction of complex information processing systems. Moreover, they are used in a wide variety of applications, in industries as well as in communication, robotics, process control, and many other fields.

Logic Gates

Logic gates are operations used in digital electronics to combine two or more simple or atomic propositions and obtain a more complex new proposition. The main logic gates are conjunction, disjunction, negation, implication, and equivalence.

Conjunction is a logic gate denoted by the symbol «Λ» and read as «AND». This operation is used to combine two simple propositions, and it results in a new proposition that is true only if both simple propositions are true. For example, if A is the proposition "Juan is tall," and B is the proposition "Ana is intelligent," then the compound proposition A Λ B would be "Juan is tall and Ana is intelligent."

Disjunction is a logic gate denoted by the symbol "V" and read as "OR." This operation is used to combine two simple propositions, and it results in a new proposition that is true if at least one of the simple propositions is true. For example, if A is the proposition "Juan is tall," and B is the proposition "Ana is intelligent," then the compound proposition A V B would be "Juan is tall or Ana is intelligent."

Negation is a logic gate denoted by the symbol "~" and read as "NOT." This operation is used to negate a simple proposition and results in a new proposition that is true if the simple proposition is false, and vice versa. For example, if A is the proposition "Juan is tall," then the compound proposition ~A would be "Juan is not tall."

Implication is a logic gate denoted by the symbol "→" and read as "IF...THEN." This operation is used to relate two propositions and results in a new proposition that is false only if the antecedent proposition is true and the consequent proposition is false. For example, if A is the proposition "Juan is tall," and B is the proposition "Ana is intelligent," then the compound proposition A → B would be "If Juan is tall, then Ana is intelligent."

Equivalence is a logic gate denoted by the symbol "↔" and read as "IF, AND ONLY IF." This operation is used to relate two propositions and results in a new proposition that is true if both propositions are true or both are false. For example, if A is the proposition "Juan is tall," and B is the proposition "Ana is intelligent," then the compound proposition A ↔ B would be "Juan is tall if, and only if, Ana is intelligent."

In Figure 5-3, the most common logic gates are shown, along with their symbols and functions.

Name	Symbol	Function
AND	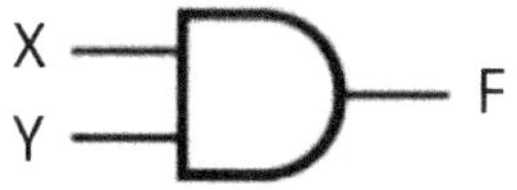	$F = X * Y$
NAND	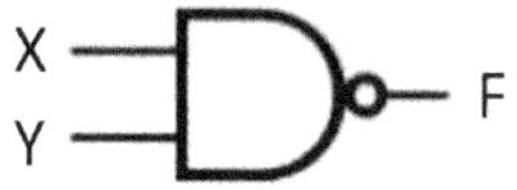	$F = \overline{X * Y}$
OR	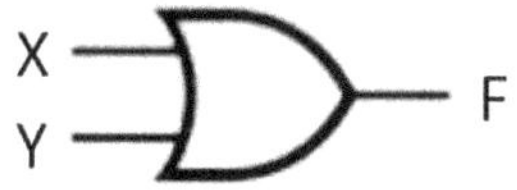	$F = X + Y$
NOR	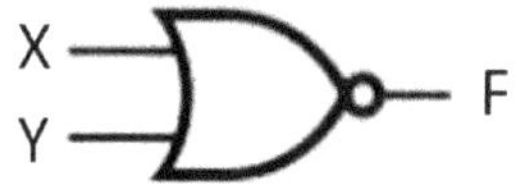	$F = \overline{X + Y}$
NOT	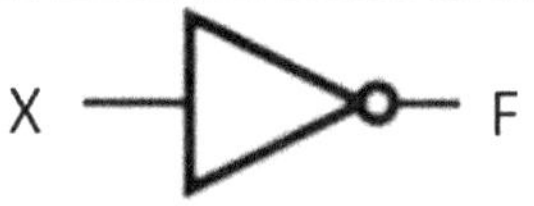	$F = \overline{X}$
XOR	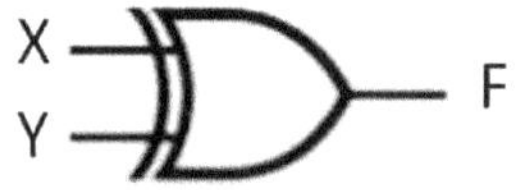	$F = X\overline{Y} + \overline{X}Y$
XNOR	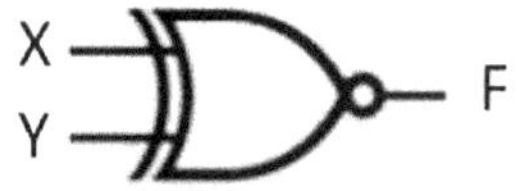	$F = XY + \overline{X}\,\overline{Y}$

Figure 5-3. Most commonly used logic gates

Flip-flops

Flip-flops are basic electronic circuits used in digital electronics to store information or data. These circuits are designed to change state (from 0 to 1 or vice versa) in response to a clock signal.

There are several types of flip-flops, but the most common ones are SR flip-flop, D flip-flop, JK flip-flop, and T flip-flop. In Figure 5-4, their respective circuits can be appreciated:

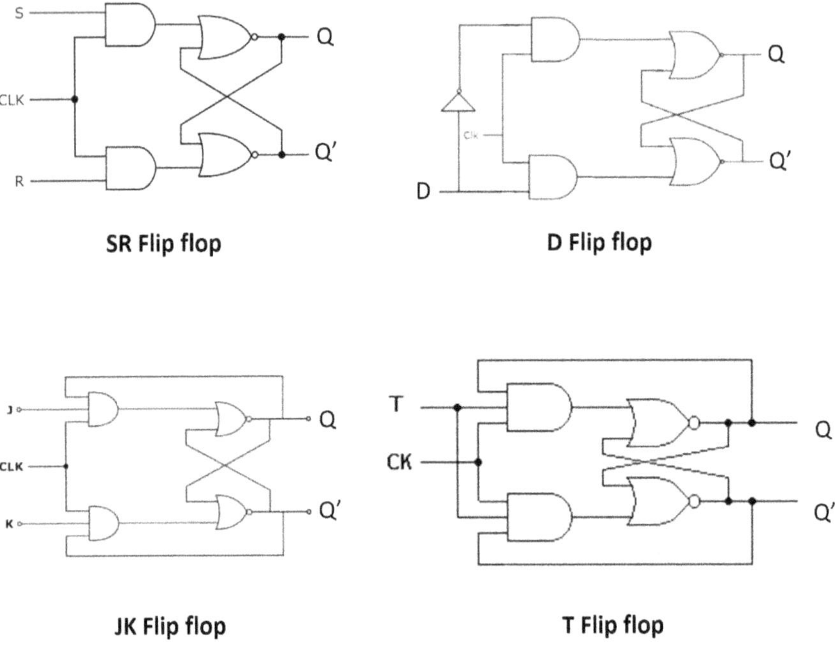

Figure 5-4. Most common flip-flops

The SR flip-flop is a circuit that has two inputs (S and R) and two outputs (Q and Q'). When the S input is activated, the Q output changes to 1, while the Q' output changes to 0. On the other hand, when the R input is activated, the Q output changes to 0, while the Q' output changes to 1. If both inputs are activated simultaneously, it results in a race condition that can generate an undesirable state.

The D flip-flop is a circuit that has a single input (D) and two outputs (Q and Q'). When the D input is activated, the Q output changes to 1, while the Q' output changes to 0. On the other hand, when the D input is deactivated, the Q output changes to 0, while the Q' output changes to 1.

The JK flip-flop is a circuit that has two inputs (J and K) and two outputs (Q and Q'). When the J input is activated and the K input is deactivated, the Q output changes to 1, while the Q' output changes to 0. On the other hand, when the K input is activated and the J input is deactivated, the Q output changes to 0, while the Q' output changes to 1. If both inputs are activated simultaneously, it results in a toggle condition that changes the state of the flip-flop.

The T flip-flop is a circuit that has a single input (T) and two outputs (Q and Q'). When the T input is activated, the Q output changes to the opposite state it had previously. That is, if Q was 1, it becomes 0, and if Q was 0, it becomes 1. The Q' output is the complementary output of Q.

Flip-flops are used in multiple applications in digital electronics, such as in counters, shift registers, memories, and in the implementation of sequential circuits.

Registers

A digital register is an electronic circuit used to store a limited amount of information in a data processing system. They are used to temporarily store data during processing and to transfer data between different parts of the system.

Digital registers are composed of a set of flip-flops that can store one or more bits of information. Each flip-flop stores a single bit of information and is controlled by clock and input/

output signals. The clock signal is a timing signal that controls when the content of the register is updated. The input/output signal is used to write or read data from the register.

There are different types of digital registers, such as shift registers, as shown in Figure 5-5, parallel load registers, feedback shift registers, universal shift registers, among others. Each type of register is adapted to different data storage needs and requirements.

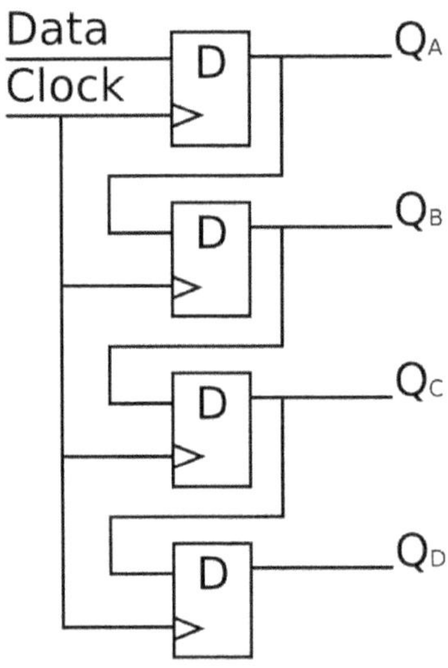

Figure 5-5. Shift register

Digital registers are used in a wide variety of applications in electronics and computer science. For example, in a computer processor, they are used to temporarily store data and memory addresses during processing. In a communication system, they are used to store data while transmitting or receiving information. In a control system, registers are used to store states and feedback values.

Counters

A digital counter is an electronic circuit used to count events or pulses in a digital system. They are composed of flip-flops and combinational circuits that perform the counting function and control the flow of data. Digital counters are used in a wide variety of applications in electronics and computer science, such as in measurement systems, signal processing, control and automation systems, among others.

There are different types of digital counters, including asynchronous counters, synchronous counters, and modulo-n counters. Asynchronous counters, also known as ripple counters, are the simplest and most economical, but their speed and precision are limited since the flip-flops are activated sequentially. Synchronous counters, on the other hand, use a common clock signal for all flip-flops, allowing for higher speed and precision. Modulo-n counters, also known as Johnson counters, use flip-flops connected in series, with one input and one output that produce a repetitive sequence of states.

Digital counters are used to count events or pulses in a digital system, such as the number of clock cycles, the number of input pulses, or the number of times a signal is activated. Counters can also be used to generate timing signals, such as clocks and timers. Moreover, counters can be cascaded to count large numbers of events or to implement complex logical functions. In Figure 5-6, the circuit corresponding to a clock is shown.

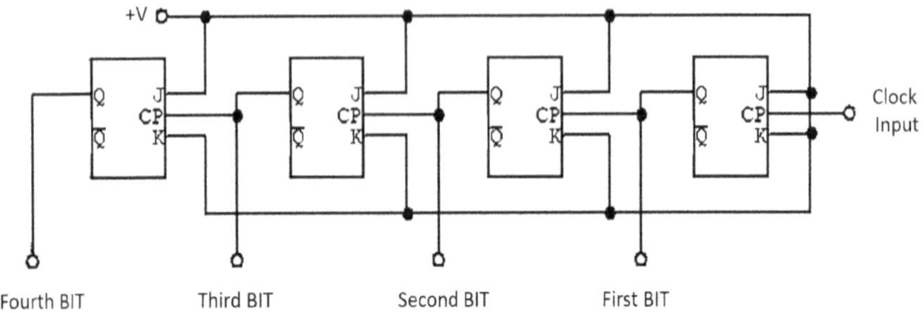

Figure 5-6. 4-Bit asynchronous counter

Review Questionnaire

Answer the question:

1. What is a binary number system and how does it work?

2. What is an isomorphism, and how is it applied in digital electronics?

3. What are logic gates, and what are their main functions?

4. What is a flip-flop, and what is its use in digital electronics?

5. What is a counter, and what is its main application in digital electronics?

Select the correct option:

6. What is the main function of Boolean algebra in digital electronics?

    e. Simplify logic circuits
    f. Convert binary numbers to decimal numbers
    g. Perform complex mathematical operations
    h. Generate audio and video signals

7. What is a flip-flop, and what is its use in digital electronics?
    a. A logic gate used to add two binary numbers
    b. A device that allows storing a bit of information and is used to build registers and counters
    c. A circuit that converts analog signals to digital signals

   d.   A device used to generate clock signals

8.   What are logic gates, and what are their main functions?
   a.   Devices used to process analog and digital signals
   b.   Circuits that convert analog signals to digital signals
   c.   Basic elements of digital logic that perform logical operations like AND, OR, and NOT
   d.   Devices used to store information in RAM

9.   What is a register, and what is its function in digital electronics?
   a.   A device used to generate clock signals
   b.   A circuit that converts analog signals to digital signals
   c.   A storage element that can store multiple bits of information and is used to maintain the state of a system
   d.   A device used to process digital signals into analog signals

10.   What is a counter, and what is its main application in digital electronics?
   a.   A circuit that converts analog signals to digital signals
   b.   A device used to process analog and digital signals
   c.   A storage element used to maintain the state of a system
   d.   A circuit that counts the number of events that occur and is used in applications like timers and pulse counters

Respond whether the statement is true or false:

11.   Logic circuits can be constructed using logic gates.
   a.   True
   b.   False

12.   An isomorphism is a relationship that exists between two circuits that have different structures but the same functionality.
   a.   True
   b.   False

13.   Flip-flops are storage elements used to maintain the state of a system.
   a.   True
   b.   False

14.   Propositional logic is a branch of mathematics used in digital electronics to perform complex mathematical operations.
   a.   True

b.   False

15. Boolean algebra is a set of mathematical rules and principles used to simplify and analyze logic circuits.
    a.   True
    b.   False

Complete the statement with the correct word or expression:

16. The _____________ is a branch of mathematics used in digital electronics to perform logical operations and simplify circuits.

17. An _____________ is a relationship that exists between two circuits that have different structures but the same functionality.

18. _____________ are storage elements that are used to maintain the state of a system in digital electronics.

19. _____________ are circuits that are used to perform logical operations in digital electronics, such as AND, OR, and NOT.

20. A _____________ is a circuit that counts the number of events that occur and is used in applications like timers and pulse counters in digital electronics.

6. MAGNETISM

Introduction

Magnetism is the property of certain materials, like magnets, that allows them to generate a magnetic field around them, which can attract or repel other magnetic materials.

Moreover, magnetic phenomena refer to the interaction between magnetic fields and magnetic materials. The latter are materials that possess magnetic properties, meaning they can generate a magnetic field or be attracted by an external magnetic field.

The magnetic field is an invisible force that surrounds a magnet or a conductor through which an electric current flows. This magnetic force can be attracted or repelled by other magnetic objects, depending on the orientation of their own magnetic poles.

One of the most common magnetic phenomena is the attraction or repulsion of objects with the same property. When two magnetic objects approach each other, their opposite poles attract, while their like poles repel.

Another important magnetic phenomenon is magnetic induction, which refers to the creation of a magnetic field in an object due to the presence of an external magnetic field. When an object is placed within a magnetic field, an electric current is generated in the object, which, in turn, generates a magnetic field. This phenomenon is used in electric generators and transformers, as we will see later.

There is also magnetization, which refers to the alignment of atoms within a magnetic material to create a permanent magnetic field. When a magnetic material is placed within an external magnetic field, the atoms inside the material align with the external field, creating a permanent magnetic field in the material. This is used in the production of permanent magnets.

Below are some basic concepts related to magnetism:

1. **Magnet**: a material that produces a magnetic field around it. It can be natural, like magnetite, or artificial, like a neodymium magnet. A magnet has two poles, as shown in Figure 6-1, called north and south, which generate opposite magnetic fields.

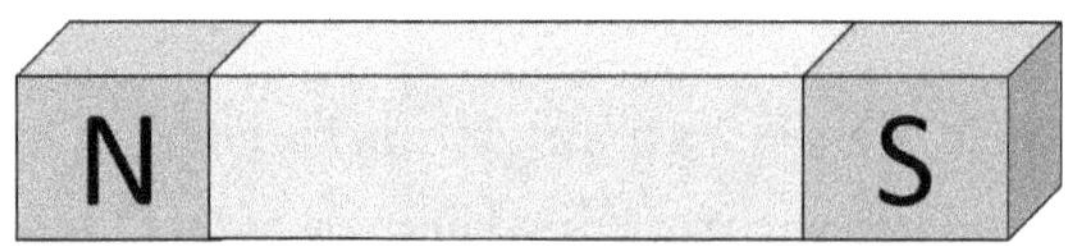

Figure 6-1. Representation of a magnet

2. **Magnetic Field**: the region of space in which a magnet

or an electric current generates a magnetic force. The magnetic field is represented by lines, as shown in Figure 6-2, indicating the direction and intensity of the magnetic field.

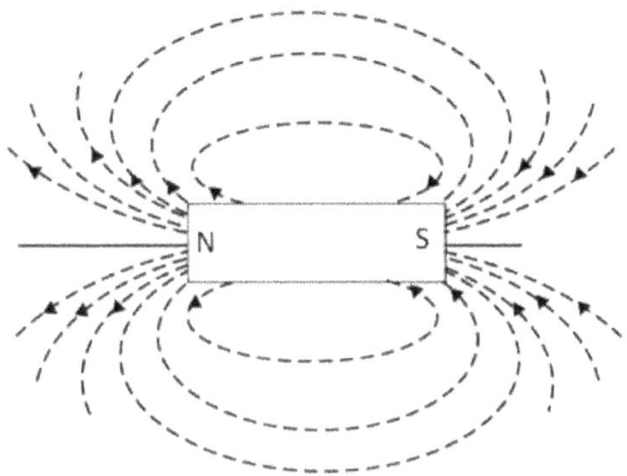

Figure 6-2. Magnetic field around a magnet

3. **Magnetic Force**: the force that occurs between two magnetic materials interacting due to their magnetic fields. The magnetic force depends on the intensity of the field and the distance between the materials.

4. **Magnetization**: the process by which the magnetic poles of a material align to generate a magnetic field in a specific direction. Magnetization can be natural or induced by an external magnetic field.

5. **Ferromagnetism**: the property of certain materials, such as iron, nickel, and cobalt, to be strongly attracted to a magnet and easily magnetizable.

6. **Paramagnetism**: the property of certain materials, like aluminum and titanium, to be weakly attracted to a magnet and capable of temporary magnetization in the presence of an external magnetic field.

7. **Diamagnetism**: the property of certain materials, like copper and gold, to be repelled by a magnet and have a negative magnetic susceptibility.

The study of magnetism is essential in physics and engineering and has many practical applications in industry, medicine, and technology.

Laws Governing Magnetic Fields

Magnetic fields are invisible, but they can be detected using instruments like a compass, which indicates the direction of the magnetic field where it is located. Additionally, magnetic fields can be measured with magnetometers, which determine the intensity of the field at a defined point.

Recall that magnetic fields are generated by the presence of moving electric charges, such as electrons in a current-carrying wire, or by the orientation of the poles in a magnetic material, like a magnet.

The magnetic force exerted on a moving electric charge in a magnetic field is calculated using the *Lorentz law*:

Ley de Lorentz (o «ley de la mano derecha»)

The magnetic force is perpendicular to both the direction of the charge's motion and the magnetic field.

Magnetic fields are divided into two main types: static and alternating. The former have a constant direction and intensity, while the latter vary in direction and intensity over time. In Figure 6-3, you can appreciate the application of the right-hand rule, which is the same as the Lorentz law, to determine the direction of a magnetic field.

Magnetic fields have many practical applications: for example, in industry, for metal separation; in medicine, for magnetic resonance imaging (MRI); and in technology, for reading and writing data on hard disks and credit cards.

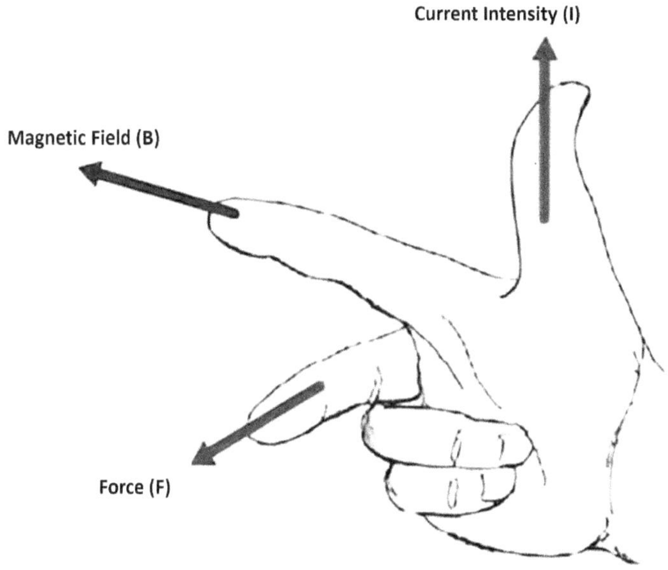

Figure 6-3. Lorentz's or right-hand rule

An example of the Lorentz law in action is the operation of an electric motor. This device works through the interaction between a magnetic field and an electric current in a conductor. The current, which is a flow of moving charges, interacts with the magnetic field generated by the magnets inside the motor. According to the Lorentz law, the Lorentz force acts on the electrons of the electric current, causing them to deviate from their original path and generate circular motion. This circular motion is what makes the motor rotate and produce mechanical work.

Faraday's Law

The variation of magnetic flux through a closed surface induces an electromotive force (EMF) in an electric circuit situated on that surface.

In other words, when the magnetic field passing through a coil changes, an electric current is generated in the coil, as shown in Figure 6-4.

Faraday's Law can be expressed mathematically using the following formula:

$$\epsilon = -\frac{d\theta}{dt}$$

Where:

$\mathcal{E}$, is the induced electromotive force in the electric circuit
$d\theta$, is the change in magnetic flux through the closed surface
dt, is the change in time

Let's consider a coil in a variable magnetic field, meaning the magnetic field passing through the coil is changing over time. In this case, according to Faraday's Law, an electromotive force is induced in the coil, producing an electric current in the same direction that opposes the original change in the magnetic field. In other words, if the magnetic field is increasing, the electric current in the coil opposes this increase in variation.

Faraday's Law is fundamental in physics and has significant applications in electricity generation, electromagnetism, and technology in general. It allows us to understand how electric currents are generated from changing magnetic fields and how these phenomena can be harnessed to generate electrical energy. Moreover, this law forms the basis of electric generator technology, where mechanical energy is converted into electrical energy through the variation of magnetic flux in a coil.

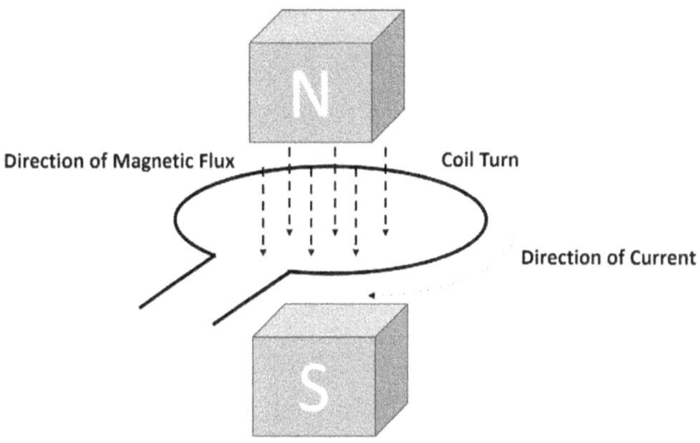

Figure 6-4. Faraday's law and Lenz's law

Lenz's Law

Any change in the magnetic field passing through a coil induces an electromotive force in the coil, which opposes the original change.

Lenz's Law can be expressed mathematically using the following formula:

$$\epsilon = -N\frac{d\theta}{dt}$$

Where:

$\mathcal{E}$, is the induced electromotive force in the coil

$N,$ is the number of turns in the coil
$d\theta,$ is the magnetic flux through the coil
$dt,$ is the change in time.

As an example of the application of Lenz's Law, let's consider a coil situated near a magnet, and the magnet approaches or moves away from the coil. In this case, a change in the magnetic field passing through the coil occurs. According to Lenz's Law, this change induces an electromotive force in the coil, generating an electric current in the same direction but opposing the original change. In other words, if the magnet approaches the coil, the generated current in the coil opposes this motion, and if the magnet moves away from the coil, the generated current in the coil opposes this movement.

Lenz's Law has significant applications in electricity generation, electromagnetism, and technology in general. In hydroelectric power plants, for example, the interaction between the magnetic field and the rotor induces an electric current in its conductors, producing electrical energy. In radio antennas, Lenz's Law is applied in the transmission and reception of radio signals. The variation of an electric field generates a magnetic field, which induces a current in a receiving antenna.

Magnetic Circuits

A magnetic circuit is a closed path of magnetic materials that allows the circulation of a magnetic field. The magnetic circuit consists of a magnetic core (usually made of iron or steel) and a coil wound around the core. The electric current flowing through the coil creates a magnetic field that induces magnetic flux in the core.

Transformers and Electric Motors are two common applications of magnetic circuits.

Transformer

A transformer consists of two insulated wire coils, called windings, placed close to each other but not touching. The winding connected to the power source is called the primary, and the other is the secondary, as shown in Figure 6-5.

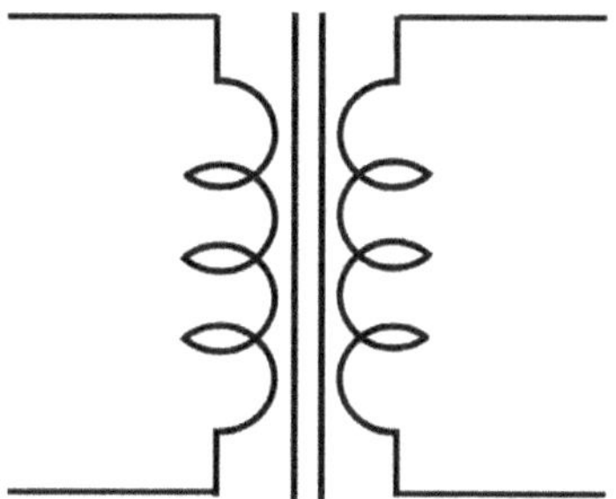

Figure 6-5. Symbol of a transformer

When electric current flows through the primary winding, it creates a magnetic field in the magnetic core. This magnetic field induces an electric current in the secondary winding, which, in turn, creates a voltage in the secondary winding. The ratio of the number of turns in the primary and secondary windings determines the voltage transformation ratio, as shown in Figure 6-6, and in the following equation:

$$\frac{V_p}{V_s} = \frac{I_s}{I_p} = \frac{N_p}{N_s}$$

Where:

Vp, is the voltage of the primary
Vs, is the voltage of the secondary

Ip, is the current of the primary
Is, is the current of the secondary
Np, is the number of turns in the primary winding
Ns, is the number of turns in the secondary winding

The operation of the transformer is based on the principles of electromagnetic induction seen earlier. When an alternating current (AC) is applied to the primary winding, a magnetic field is produced that passes through the transformer's core, inducing a current in the secondary winding.

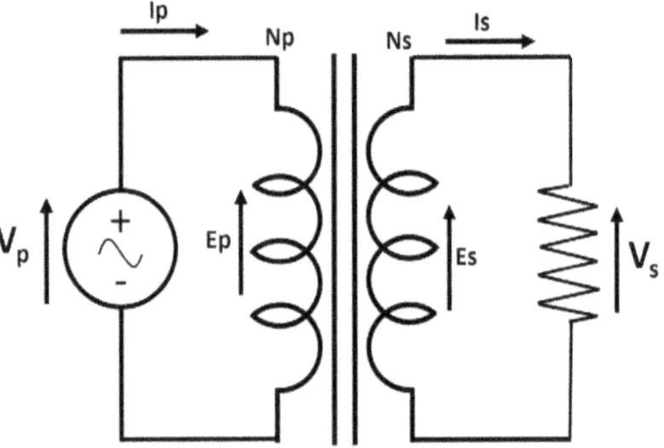

Figure 6-6. Relationships in a transformer

It is essential to note that in an ideal transformer, the power in the primary and secondary windings is always the same, in accordance with the law of conservation of energy, which states that energy cannot be created or destroyed; it can only be transformed.

In a real transformer, however, there are losses due to the resistance of materials and the dissipation of energy in the form of heat. Therefore, the power in the secondary will be slightly less than the power in the primary.

Distribution transformers are used to reduce the high voltage at which energy is generated to a lower voltage, which is used in households and businesses. Transformers are also used in electronic devices to provide isolation and reduce electrical noise.

Motor

An electric motor is a device that converts electrical energy into mechanical energy. It consists of a rotor (moving part) and a stator (fixed part), which are located within a casing.

The rotor is composed of a shaft and a set of coils (windings) wound around it. These coils are supplied with electric current, creating a magnetic field in the rotor. The stator, on the other hand, is made up of a set of permanent magnets or coils that generate a fixed magnetic field.

When electric current flows through the coils of the rotor, a magnetic field is produced that interacts with the magnetic field of the stator. This rotating magnetic field generates a torque that makes the *rotor* spin. The rotational speed of the rotor depends on the frequency of the electric current and the motor's design.

There are different types of electric motors, such as direct current (DC) motors, alternating current (AC) motors, and stepper motors. DC motors are commonly used in applications that require precise speed control, such as in robots and electric vehicles. AC motors, on the other hand, are more common in industrial applications, such as pumps and fans.

In stepper motors, the rotor rotates in discrete steps rather than continuously. These motors are used in applications that require high positioning precision, such as 3D printers and laser cutting machines.

Electromagnetic Waves

Electromagnetic waves are a type of wave that propagates through space through the interaction of oscillating electric and magnetic fields. These waves are generated from the vibration of electric charges and propagate through vacuum and various material mediums.

Electromagnetic waves have a set of properties that make them unique and very useful in different fields of science and technology. Some of the most important properties of electromagnetic waves are:

8. **Frequency**: Refers to the number of complete cycles of oscillation it performs per unit of time. The unit of frequency is the Hertz (Hz). Electromagnetic waves are classified based on their frequency into different categories, such as radio waves, microwaves, visible light, X-rays, etc.

9. **Wavelength**: Refers to the distance between two equivalent points of the wave, such as two consecutive crests or two consecutive troughs. The unit of wavelength is the meter (m).

10. **Propagation Speed**: Refers to how fast the wave travels through space or the material medium it is in. The propagation speed of electromagnetic waves in a vacuum is 300,000 kilometers per second (speed of light), denoted by the letter "C".

11. **Polarization**: Refers to the direction of oscillation of the electric and magnetic fields that compose it. Electromagnetic waves can be polarized in different ways, such as li-

near, circular, or elliptical.

12. **Interference**: Refers to the phenomenon that occurs when two or more waves overlap at a point in space. Depending on the relative phase of the waves, they can interfere constructively or destructively, producing either reinforcement or cancellation of the resulting wave.

Electromagnetism

Electromagnetism is a branch of physics that studies the relationship between electric and magnetic fields and their effects on moving electric charges. In other words, it seeks to understand how electricity and magnetism are related and how they interact with each other.

The relationship between electromagnetism and magnetism arises because a magnetic field is generated when there are moving electric charges. As we have seen, when an electric current flows through a conductor, a magnetic field is produced around it. Additionally, if a conductor is placed in a magnetic field, an electric current is induced in it, according to Faraday's law.

The theory of electromagnetism is described by Maxwell's equations, which establish the relationship between electric and magnetic fields and how they propagate through space. These equations also describe how electric charges interact with electric and magnetic fields and how electromagnetic waves, such as light, are generated.

Review Questionnaire

Select the correct option:

1. What is magnetism?

    a. The ability of an object to attract metals.
    b. The ability of an object to repel metals.
    c. The ability of an object to attract or repel metals.

2. What is a magnet?

    a. An object that attracts metals.
    b. An object that repels metals.
    c. An object that attracts or repels metals.

3. What is the magnetic field?

    a. A region of space where a magnetic force is exerted.
    b. The distance between two magnets.
    c. The magnetic force that one magnet exerts on another.

4. What is magnetic force?

    a. The force that one magnet exerts on another magnet.
    b. The force that a magnetic field exerts on a moving electric charge.
    c. The force that a magnetic field exerts on a stationary electric charge.

5. What is magnetization?

    a. The process of creating a magnetic field in an object.
    b. The process of creating a magnet from a non-magnetic object.
    c. The process of creating an electric charge in an object.

Answer whether the statement is true or false:

6. Ferromagnetism is the property of certain materials to be permanently magnetized.

    a. True
    b. False

7. The Lorentz Law (or the right-hand rule) determines the direction of the magnetic force on a moving electric charge in a magnetic field.

    a.   True
    b.   False

8. A transformer is a device that transforms mechanical energy into electrical energy.

    a.   True
    b.   False

9. A transformer is a device that transforms voltage and electric current in a circuit to a different level.

    a.   True
    b.   False

10. Lenz's law states that the induced electric current in a circuit is proportional to the magnetic force on the moving electric charge.

    a.   True
    b.   False

Answer the question:

11. What is electromagnetism?

12. What are electromagnetic waves?

13. What are the laws that govern magnetic fields?

14. What does Faraday's law establish?

15. What is a transformer used for?

Complete the statement with the correct word or expression:

16. The _________ is a region of space where a magnetic force is exerted on a magnetic charge at rest.

17. The transformer functions based on the principle of ________, which es-

tablishes that the ratio between the number of turns in the primary and secondary determines the relationship between voltage and current on both sides of the transformer.

18. What are the particles that move in electrical conductors to produce a magnetic field?

19. Which law describes electromagnetic induction?

20. Which law describes the direction of the induced current in a coil when the magnetic field passing through it changes?.

7. ELECTRIC POWER SYSTEMS

Definition

Electric Power Systems (EPS) are interconnected networks of generation, transformation, transmission, and distribution of electric energy. They are critical processes for modern infrastructure (industries, businesses, and households worldwide) in the current, voltage, power, and energy levels they require.

In the previous chapters, we have already mentioned the electrical equipment and components involved in the process of generation, transformation, transmission, and distribution of electric energy, such as turbines, generators, transformers, among other components that are less visible but always very relevant, such as coils, capacitors, etc. Now we are going to integrate them into an integral process, based on Figure 7-1:

1. **Generation**: as we have seen, electric energy is produced

in a power plant through the transformation of a primary energy source (in this case, hydroelectric). The electric energy generated in the power plant is at medium voltage, so a transformation is required to transport it over long distances.

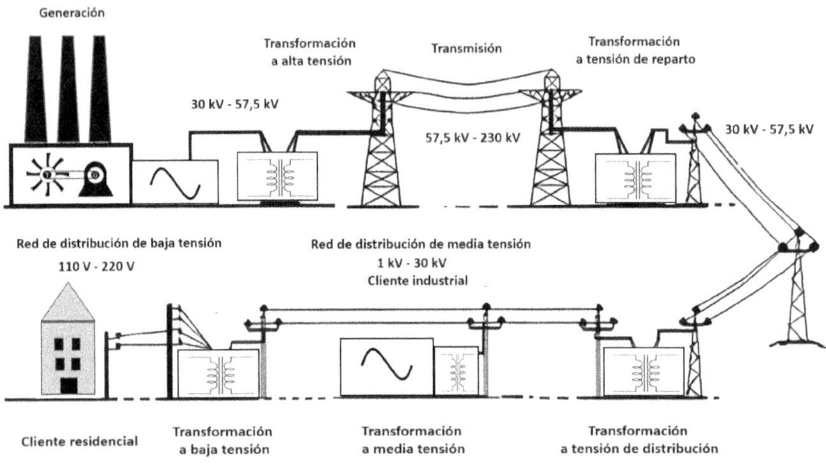

Figure 7-1. Process of generation, transformation, transmission, and distribution of electrical energy

2. **Transformation to high voltage**: the electric energy generated at medium voltage is passed through step-up transformers. This transformation allows electric energy to be efficiently transported with minimal losses.

3. **Transmission**: it is carried out through high-capacity transmission lines over long distances. These lines are built with sturdy materials to withstand environmental conditions and electric load.

4. **Transformation to distribution voltage**: the electric energy arrives at the substation, where it is reduced back

to medium voltage through power transformers. The substation may also include other control elements, such as switches and protective devices, to ensure the stability and safety of the electric system.

5. **Distribution network**: from the substation, electric energy is distributed through the distribution network, a network of medium-voltage lines that deliver electric energy to distribution transformers located at different points in the area to be served.

6. **Transformation to distribution voltage**: electric energy is transformed to different voltage levels as it gets closer to the end customer. Generally, it is transformed to levels between 1 and 30 kV for distribution at medium voltage.

7. **Distribution at medium voltage to industrial customers:** electric energy is distributed through medium-voltage lines to industrial customers, where it is used to power motors and electrical equipment.

8. **Low-voltage transformation center**: for distribution to residential customers, electric energy is transformed at a low-voltage transformation center, which reduces the voltage to levels between 110 and 220 volts.

9. **Delivery to residential customers**: electric energy is delivered to residential customers through low-voltage lines.

Electric power systems must be designed and operated safely and efficiently to ensure the reliability of the electric supply. This includes managing electric load, balancing generation and demand, protecting against faults, and planning for system expansion.

Safety is a critical consideration in electric power systems (which we will delve into later), as system failures can have serious consequences, such as widespread blackouts, fires, and injuries to people and animals. For this reason, electric power systems are designed to protect themselves and their equipment from electrical faults.

Transmission and Distribution Systems

Transmission and distribution systems of electric energy are an essential part of electric power systems. They are electrical networks, consisting of cables supported on steel lattice towers, designed to transport large amounts of energy (at high, medium, and low voltage) from generation points to end consumers, such as households, businesses, and industries.

Transmission is carried out through high-voltage lines, which can transport large amounts of electric energy over long distances to distribution substations. These lines are designed to minimize energy losses due to low cable resistance, which means that most lines are made of aluminum or copper, materials with high electrical conductivity. The towers or poles serve to prevent contact between the cables and vegetation, as well as to keep them at a safe distance from the population. Figure 7-2 shows two transmission towers.

Distribution, on the other hand, is carried out through medium and low voltage lines that connect electrical substations to end consumers. These distribution lines are designed to carry a lower amount of electric energy (compared to high-voltage lines) over shorter distances. Distribution also requires transformers that reduce the voltage of the transmission lines to safe levels for consumers.

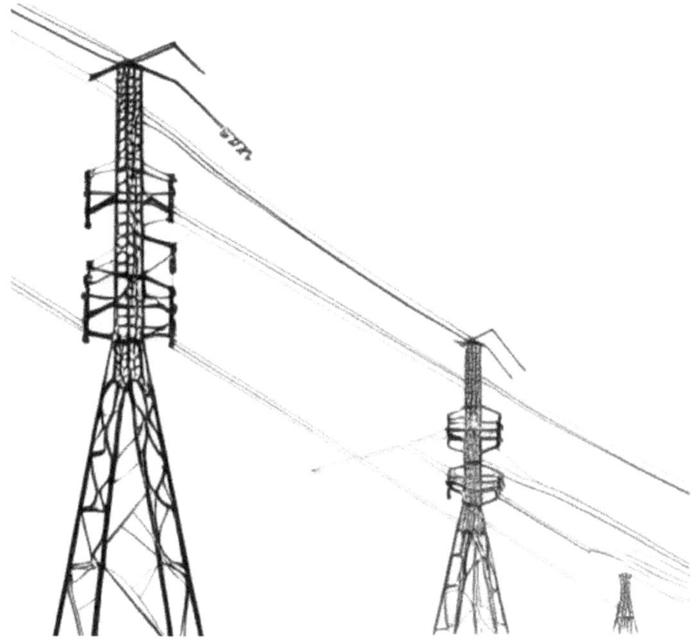

Figure 7-2. Power transmission towers

In addition to transmission and distribution lines, electric power systems also have transformer substations, which are used to connect different transmission lines and to change the voltage of electric energy at connection points. Transformer substations also have protection and control systems to ensure the stability and safety of the electric system.

To change the voltage, transformers are used, which increase or decrease its level to minimize energy losses and to adapt the voltage to the levels required by consumption center.

The efficiency of the transmission and distribution system

is key to the reliability of the electric supply. To ensure a safe and constant supply of electric energy, transmission and distribution systems are designed with redundancies and protection measures against faults, such as automatic switches and overload protection devices, which we will explore further.

Three-Phase Systems

Three-phase is an electrical distribution system that uses three sinusoidal alternating current phases with the same frequency and amplitude, but displaced from each other by an angle of 120 degrees. This type of electric system is used in most electric energy distribution networks worldwide, including high and low-voltage networks.

Electric energy is generated in a power plant in the form of three-phase alternating current, which is then transmitted through high-voltage transmission lines. In the transformer substations, it is reduced to lower voltage levels and distributed to end consumers.

There are several advantages to using three-phase systems in the distribution of electric energy. One of them is greater efficiency in transmission and distribution. Because the alternating currents in a three-phase system are 120 degrees out of phase, the individual currents sum up to produce a resultant current with a constant amplitude. This reduces energy losses due to cable resistance, which in turn lowers the cost of energy transmission and distribution.

Another advantage of the three-phase system is that it is easier to balance compared to a single-phase system (which we will discuss next). In a single-phase system, electrical load must be distributed evenly among different circuits to avoid imbalances. In a three-phase system, imbalances are automatically reduced due to the nature of the phase-displaced alternating currents.

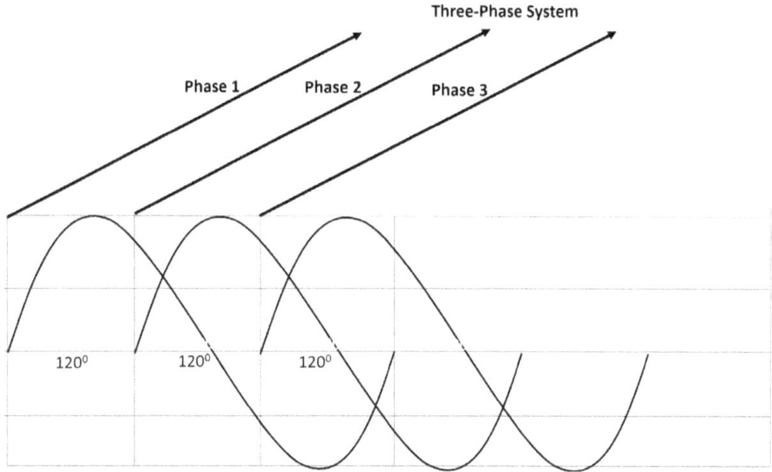

Figure 7-3. Three-phase system

Single-Phase Systems

Single-phase is an electrical distribution system that carries a single phase of sinusoidal alternating current, which varies in magnitude and direction over time, as shown in Figure 7-4. It is used in low-power applications and for small-scale energy distribution, such as in homes and small businesses.

In a single-phase system, electric energy is generated in a power plant as single-phase alternating current, and then transmitted through low-voltage transmission lines. In the transformer substations, the energy is reduced to lower levels for distribution to end consumers.

Unlike three-phase systems, single-phase systems have several disadvantages. The main one is that due to the varying magnitude and direction of the current over time, there is a fluctuation in the power delivered to the load, which can affect the efficiency of the electrical and electronic equipment using it, thus reducing its lifespan.

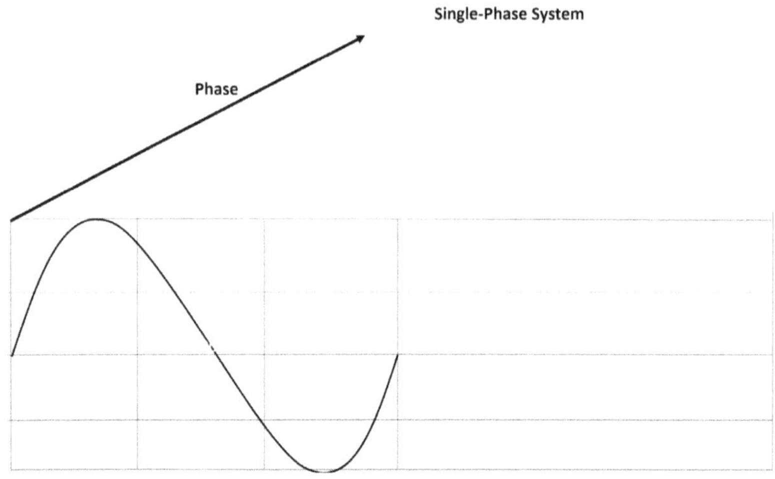

Figure 7-4. Single-phase system

Another disadvantage of this system is that it is not as efficient as the three-phase system in transmission and distribution. Since there is only one phase, the summation effect of currents cannot be utilized, increasing energy losses due to cable resistanc.

Conversion from Three-Phase to Single-Phase Power

The conversion from three-phase to single-phase power is a

common process. This is done to use electrical and electronic equipment that requires single-phase power in areas where only three-phase power is available, as shown in Figure 7-5.

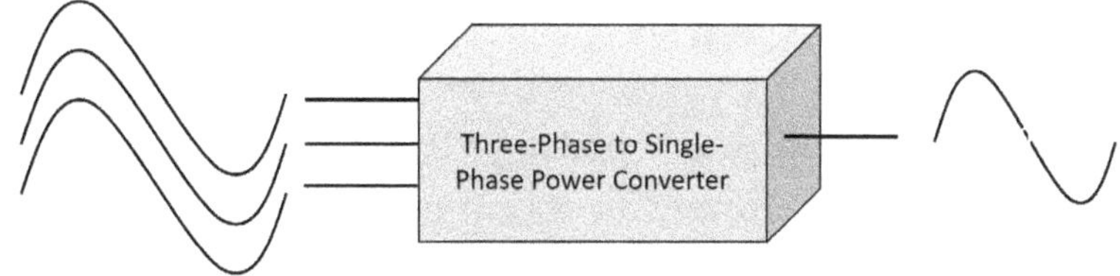

Figure 7-5. Three-phase to single-phase energy converter

There are several ways to achieve this conversion. One of the most common methods is by using a three-phase transformer with a single-phase secondary connection. The transformer is used to reduce the voltage of the three phases and provide a single-phase secondary connection that supplies power to the electrical and electronic equipment requiring single-phase power.

Another common way to convert three-phase to single-phase power is by using electronic devices such as frequency converters. These devices perform the conversion by using electronic circuits that control the phase and frequency of the electric current. This is especially useful in applications where motor speed and other electrical equipment need to be controlled.

It is essential to highlight that converting from three-phase to single-phase power can cause certain limitations in the performance of electrical and electronic equipment. Some equipment, such as motors, may lose efficiency during the conversion process due to transformation processes. Therefore, it is essential to carefully consider the limitations and performance of the equipment before opting for this power conversion.

Conversion from Single-Phase to Three-Phase Power

The conversion from single-phase to three-phase electric power is a process used to supply electrical equipment and electronic devices that require three-phase power in locations where only single-phase power is available, as shown in Figure 7-6.

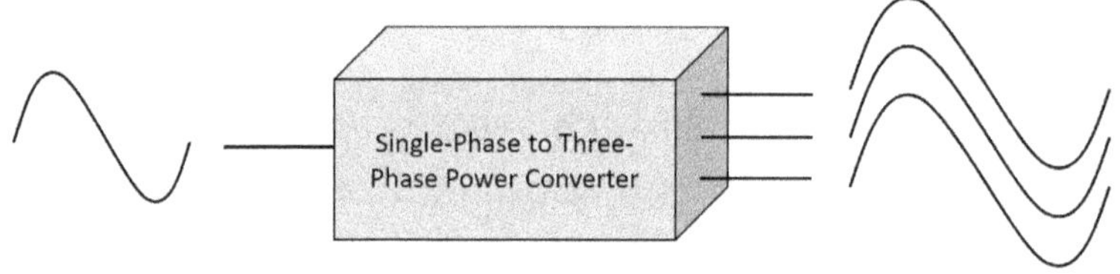

Figure 7-6. Single-phase to three-phase energy converter

The most common way to convert single-phase electric power to three-phase power is by using a static converter. These are electronic devices that control the phase and frequency of the electric current.

Another way to convert single-phase electric power to three-phase power is by using a single-phase electric motor, which cannot operate directly on three-phase power but can be used to generate a form of three-phase electric power (from single-phase electric power) using a three-phase capacitor and a switching system that controls the sequence of the electric current.

As in the previous case, it is important to mention the limitations of this type of conversion, as the performance of electrical and electronic equipment can decrease considerably. This is due to losses in the static converters or switching systems used for the conversion.

Review Questionnaire

Answer the question:

1. What is an electrical transmission and distribution system?

2. What is the difference between a three-phase system and a single-phase system?

3. What advantages does the use of three-phase systems have in electrical transmission and distribution?

4. Why is the conversion from three-phase to single-phase power and vice versa necessary?

5. What equipment is used to perform the conversion from three-phase to single-phase power and vice versa?

Select the correct option:

6. What is the definition of electrical transmission and distribution systems?
   a. Equipment used to generate electrical energy
   b. Infrastructure used to carry electrical energy from the generating station to the end consumer
   c. Tools used to measure the electrical energy consumed by a home or business
   d. Devices used to store electrical energy

7. What is the main difference between a three-phase system and a single-phase system?
   a. The number of phases
   b. The frequency of the electrical current

    c.   The voltage of the electrical current
    d.   The type of electrical current used

8. What advantage does the use of three-phase systems offer in electrical transmission and distribution?
    a.   Greater safety for personnel working in the electrical network
    b.   Higher efficiency in electrical transmission and distribution
    c.   Better quality of electrical energy supplied to the end consumer
    d.   Lower cost of equipment used in the electrical network

9. Why is the conversion from three-phase to single-phase power and vice versa necessary?
    a.   To reduce the amount of electrical energy consumed by a home or business
    b.   To adapt the electrical energy supplied by the electrical network to the needs of consumers
    c.   To increase the amount of electrical energy supplied by the electrical network
    d.   To reduce the amount of equipment required in the electrical network

10. What equipment is used to perform the conversion from three-phase to single-phase power and vice versa?
    a.   Transformers and rectifiers
    b.   Inverters and chargers
    c.   Capacitors and coils
    d.   Switches and fusess

Answer if the statement is false or true:

11. The definition of electrical transmission and distribution systems refers to the equipment used to generate electrical energy.
    a.   False
    b.   True

12. Three-phase systems have a higher frequency of electrical current than single-phase systems.
    a. False
    b. True

13. The use of three-phase systems in electrical transmission and distribution offers higher efficiency.
    a. False
    b. True

14. The conversion from three-phase to single-phase power and vice versa is necessary to reduce the number of equipment required in the electrical network.
    a. False
    b. True

15. The equipment used to perform the conversion from three-phase to single-phase power and vice versa is transformers and rectifiers.
    a. False
    b. True

Complete with the word or expression that corresponds:

16. The __________________ of electrical transmission and distribution systems refers to the infrastructure used to carry electrical energy from the generating station to the end consumer.

17. In three-phase systems, the electrical current is divided into __________________ phases.

18. Single-phase systems use a single __________________ of electrical current.

19. The conversion from three-phase to single-phase power is used to adapt the electrical energy supplied by the electrical network to the needs of __________________.

20. The conversion from single-phase to three-phase power is used to increase the amount of electrical energy supplied by the electrical network and reduce the costs of __________________.

8. ELECTRICAL SYSTEM CONTROL

Definition

Electrical system control is an important discipline of electrical engineering that deals with regulating the operation of electrical systems to ensure their efficiency, safety, and reliability. Control can be achieved using various devices and techniques, among them, those based on thyristors, which are often referred to as industrial electronics.

Thyristors

Thyristors are semiconductor devices used to control electric current in high-power circuits. They typically have three terminals: anode, cathode, and gate, as shown in Figure 8-1 (a) and (c), and are activated by applying a current pulse to the gate terminal. Once the thyristor is activated, it allows current to flow from the anode to the cathode and remains in that state until the current is interrupted through the circuit. In other words, it functions as a short circuit (in the ON state) or as an open circuit (in the OFF state), as indicated in Figure 8-1 (b).

Doping

A very important process used in the manufacturing of thyristors is *doping*. This involves adding controlled impurities to the semiconductor materials that make up the device. These impurities, also known as *dopants*, are atoms of other chemical elements that modify the electrical properties of the semiconductor. The type and amount of dopants added to the different layers of the thyristor determine its conductivity, leakage current, breakdown voltage, and holding current.

1. **Conductivity**: It is the ability of the element to conduct electric current.

2. **Leakage current:** It is the amount of electric current that flows through the thyristor when it should be turned off.

3. **Breakdown voltage**: It is the maximum voltage that the thyristor can withstand before being damaged.

4. **Holding current**: It is the minimum current required to keep the thyristor in its ON state.

In general, two types of dopants are used in thyristor manufacturing: P-type impurities and N-type impurities. P-type impurities, such as boron, are added to the N-type region to create a junction barrier. On the other hand, N-type impurities, such as phosphorus, are added to the P-type region to increase its conductivity.

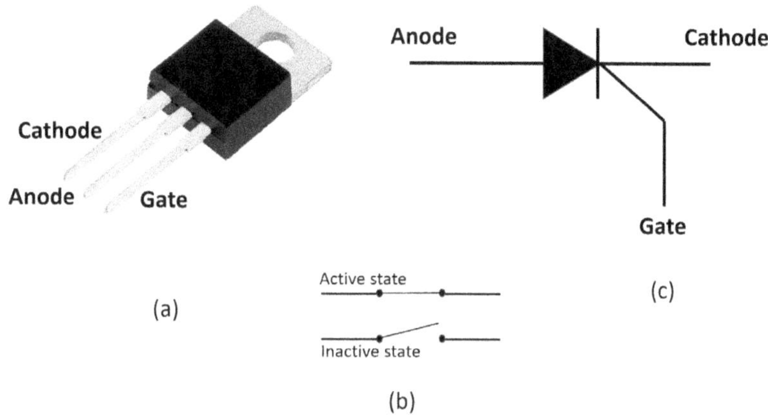

Figure 8-1 (a) Generic form of a thyristor, (b) Thyristor behavior, and (c) Symbol of a thyristor

The doping process is critical in manufacturing, not only of thyristors but also of electronic devices in general. Incorrect doping can cause device failures and affect performance. Therefore, it is important to carry out the doping process accurately and in a controlled manner to ensure the quality and reliability of the device..

Terminology

The terminology of thyristors may vary depending on the manufacturer or geographic region. Although there are some standard terms used worldwide, some manufacturers may use different terms to refer to the same thing.

For example, the term SCR is commonly used to refer to Silicon-Controlled Rectifiers, but it can also be used to refer to any type of thyristor. Similarly, the term TRIAC is often used to refer to any thyristor that can be activated in both directions of current flow, although technically it refers to a specific type of three-terminal thyristor.

Another example of variable terminology is the use of expressions like "anode" and "cathode." Some manufacturers or researchers often use "anode 1" and "anode 2" to refer to the thyristor terminals; the same occurs with the terms "gate" or "gate," used to refer to the control terminal in thyristors. Even when analyzing the layer composition of thyristors, devices with three, four, and even five layers frequently appear. Although in some countries and for some manufacturers, the terms are interchangeable and all are correct, users may find it confusing to encounter so many different terms.

To address potential difficulty with terminology, it is essential to note that, although it may vary, thyristors have defined specifications and technical characteristics that must be understood for proper use. Therefore, it is vital to consider both terminology and technical specifications and functionality when working with thyristors.

Operation of Thyristors in AC and DC

Now, the operation of thyristors varies depending on whether they are used in alternating current (AC) or direct current (DC).

In alternating current, thyristors are used as phase-controlled switches. As previously mentioned, this means they are activated during a certain period of time in each cycle of the AC signal. One application in this mode is delaying the current flow to a resistive, inductive, or capacitive load, thus controlling the amount of energy flowing through the circuit. In general, AC thyristors are used in high-power applications, such as motor control, high-intensity lighting, and power regulation in the electrical grid.

In direct current, thyristors are used as direct current switches, allowing the control of the amount of energy flowing through a circuit. In this case, they are used as switching devices to control the power supply to loads such as motors or resistors in DC circuits. They are used in both low and high-power applications, such as motor controllers, low-power lighting, and power electronics in general.

Types of Thyristors

There are different types of thyristors, each with specific characteristics that depend on the composition of their layers and the doping applied to them. Some of the most commonly used types include:

The Silicon-Controlled Rectifier (SCR)

It is a type of four-layer diode that allows the conduction of electric current in a single direction, but only when an appropriate voltage pulse is applied to the third layer, or gate. In this case, the SCR switches from a blocking state to a conducting state, allowing current to flow through the device. Figure 8-2 shows (a) the layered composition and (b) the symbol of the SCR.

The SCR is used to control the flow of electric current in alternating current (AC) circuits, as well as in power control circuits, such as voltage regulators, motor speed controllers, and temperature control systems. It is also employed in high-power electronics, such as energy converters, inverters, and rectifiers.

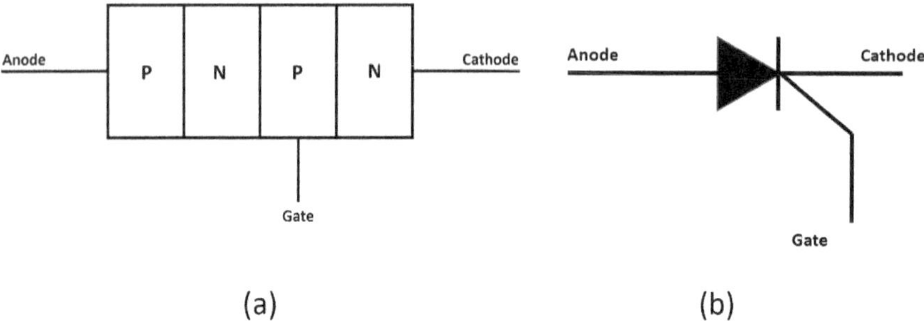

Figure 8-2.SCR layer composition

It's important to note that the SCR is a direct current (DC) device, so additional circuits are required to rectify and filter the alternating current before applying it to the SCR. Additionally, due to its unidirectional conducting nature, it is not suitable for applications that require fast current switching in both directions.

DIAC (Alternating Current Diode)

It is a semiconductor device used in power control electronic circuits to regulate the electric current flowing through alternating current circuits.

Unlike most thyristors, the DIAC has only two terminals and acts as an on-off switch for the current. It consists of two diodes connected in parallel and configured in opposite directions, as shown in Figure 8-3 (b).

As seen in Figure 8-3 (a), the DIAC has three layers and two terminals. The upper and lower layers contain both N and P materials. It is considered a PNPN structure with the same characteristics as a four-layer diode.

The DIAC's threshold voltage is the voltage from which the device conducts in both directions; if the applied voltage is lower than the threshold voltage, the device has a high impedance and does not conduct.

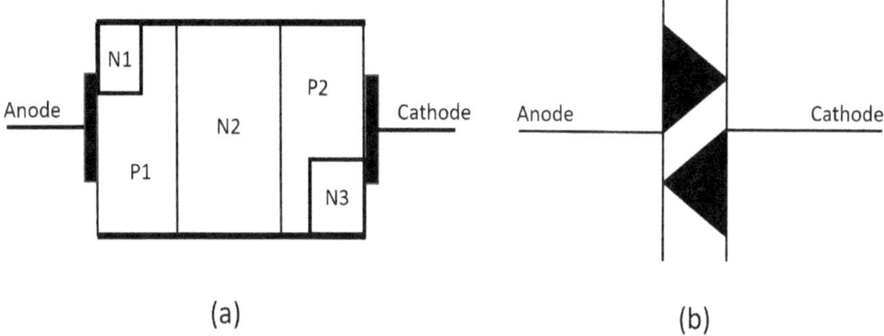

(a)   (b)

Figure 8-3 (a) Layer composition, (b) Symbol of DIAC

The DIAC is often used in thyristor triggering applications, such as the TRIAC (which we will see below). When a triggering current is applied to the gate of a TRIAC through a DIAC, the TRIAC is activated, allowing the alternating current to flow through the circuit. In this way, the DIAC is used to control the alternating current and adjust the signal intensity in a circuit.

It can be used in temperature control circuits, lighting circuits, and other electrical devices where precise current and voltage regulation are required.

TRIAC (Alternating Current Triode)

The TRIAC is similar to a DIAC but with a gate terminal. A

TRIAC can be triggered by a gate current pulse and does not require a threshold voltage to initiate conduction, like the DIAC. It can also be thought of as two SCR devices connected in parallel in opposite directions with a common terminal, which is the gate. However, unlike the SCR, the TRIAC can conduct current in either direction when activated, depending on the polarity of the voltage across its terminals. Figure 8-4 shows (a) the layered composition of the TRIAC and (b) the TRIAC symbol.

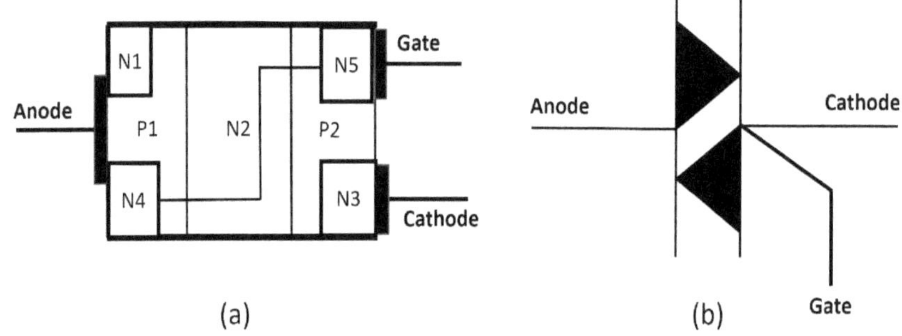

Figure 8-4 (a) Layer composition, (b) Symbol of TRIAC

The TRIAC is used for voltage regulation in power control circuits, lighting circuits, and variable-speed motors.

In lighting circuits, for example, it can be used to control the light intensity by varying the amount of current flowing through a bulb. In variable-speed motors, it is used to control the motor's speed by regulating the amount of current flowing through it.

Protection of Electrical Systems

The protection of electrical systems is an important part of the design and operation of any electrical system. It refers to safeguarding electrical equipment and systems against damage and failures that may occur due to overloads, short circuits, ground faults, among others.

This is achieved through the use of electrical protection devices, such as circuit breakers, protective relays, fuses, and disconnecting devices. These devices automatically detect and disconnect any faults or overloads in the system to prevent damage to equipment and people.

Ground faults, for example, can be dangerous for people and can cause damage to electrical equipment. Ground fault protection devices, such as differential switches, automatically detect and disconnect any ground faults in the electrical system.

The protection of electrical systems also includes protection against overvoltages and voltage fluctuations. Protection devices are used to safeguard electrical equipment against overvoltages that may occur due to lightning storms or network failures. Voltage fluctuation protection devices, such as voltage regulators, are used to protect electrical equipment against voltage fluctuations that can damage them.

It is important to highlight that the protection of electrical systems is an ongoing process that requires regular maintenance and testing to ensure that protection devices are working correctly and to ensure the reliability and efficiency of the electrical system.

Disconnection Devices

Disconnection devices are electrical components designed to

automatically disconnect electric current in case of an emergency or system failure. These devices are vital for the safety and protection of equipment and people working with them.

There are different types of disconnection devices, each designed for specific applications. Here are some common examples:

Automatic Circuit Breakers

Automatic circuit breakers, also known as thermomagnetic circuit breakers, are electrical devices used to protect electrical circuits and connected equipment from overloads and short circuits.

These breakers are different from conventional switches in that they can automatically detect overload and short circuit conditions in the circuit and quickly and safely disconnect the electric current to prevent damage or fires.

Automatic circuit breakers work through a thermal and magnetic tripping mechanism. The thermal mechanism detects temperature rise caused by overload in the circuit and automatically disconnects the circuit before the overload can damage the equipment. The magnetic mechanism, on the other hand, detects excessive current caused by a short circuit and immediately disconnects the circuit.

These breakers are widely used in electrical installations of all types, from small residential applications to large industrial and commercial facilities. They are very reliable and easy to maintain, and their ability to automatically detect and disconnect faulty circuits makes them a popular choice for electrical protection.

Protection Relays

A protection relay is an electrical device used to protect electrical equipment and systems from possible electrical faults or overloads. They are triggered when they detect an abnormal condition in the system, such as current overload, ground fault, voltage loss, among others. Once triggered, they interrupt the power supply to the protected equipment or system, avoiding potential damage or accidents.

Protection relays consist of several parts, including a magnetic core, a set of electrical contacts, and an activation mechanism. The magnetic core is used to detect electrical currents and transform them into electrical signals that can activate the relay's activation mechanism. The electrical contacts are used to interrupt the power supply to the protected equipment or system when the relay is activated.

There are different types of protection relays, each designed to protect electrical equipment and systems from different types of faults and overloads. Some examples of common protection relays include thermal overload relays, ground fault relays, phase sequence protection relays, low-voltage protection relays, among others.

Fuses

Fuses are disconnection devices used to protect electrical circuits against overloads and short circuits. When an overload or short circuit occurs, the fuse melts, disconnecting the electric current. As we saw in the electronics chapter.

Ground Fault Circuit Interrupters (GFCI)

These are used to protect against ground faults in electrical systems. If a ground fault is detected, the GFCI automatically disconnects the electric current.

Safety Switches

These are used to disconnect electric current in emergency situations, such as a fire or flood.

Air Break Switches

These devices are used in high-voltage systems to disconnect the electric current in case of a system failure.

Review Questionnaire

Answer the question:

1. What is a thyristor and what is its function in an electrical circuit?

2. What does the term "doping" mean in relation to the semiconductor materials used in thyristors?

3. What are the main types of thyristors and how do they differ?

4. How does an SCR work and what is it commonly used for?

5. What is the difference between a DIAC and a TRIAC, and what applications is each one used for?

Select the correct option:

6. What is the definition of a thyristor?

    a.   A device that controls the flow of electric current.
    b.   A device that stores electrical energy.
    c.   A device that generates electricity from solar light.

7.   What is doping in the semiconductor materials used in thyristors?
    a.   The addition of impurities to modify the electrical properties of the material.
    b.   The manufacturing process of semiconductor materials.
    c.   The measurement of electrical conductivity of semiconductor materials.

8.   What is the function of thyristors in alternating current (AC) systems?
    a.   To control the flow of current in one direction.
    b.   To control the flow of current in both directions.
    c.   To convert alternating current into direct current.

9.   What protection device is commonly used in electrical systems to protect against current overloads?
    a.   Fuses.
    b.   Air break switches.
    c.   Automatic circuit breakers.

10. What is the function of protection relays in electrical systems?
    a.   To disconnect the system in case of a failure.
    b.   To control the amount of current flowing in the system.
    c.   To protect the system against electrical discharges.

Answer if the statement is true or false:

11. Thyristors are devices that control the flow of electric current.
    a.   True
    b.   False

12. Doping is the manufacturing process of semiconductor materials used in thyristors.
    a.   True
    b.   False

13. Thyristors can only be used in alternating current systems.
    a.   True
    b.   False

14. Fuses are devices that protect electrical systems against current overloads.
    a. True
    b. False

15. Automatic circuit breakers are protection devices commonly used in electrical systems.
    a. True
    b. False

Complete with the appropriate word or expression:

16. Doping is the process of introducing impurities into a semiconductor material to modify its electrical properties, turning it into a material with __________ characteristics.

17. Automatic circuit breakers are devices that protect electrical systems against current overloads and short circuits, and they automatically activate when the electric current exceeds a certain __________.

18. The DIAC is a semiconductor device used in alternating current circuits to allow the passage of current in both directions when a specific __________ is exceeded.

19. The TRIAC is a semiconductor device used in alternating current circuits to control the flow of current by applying a __________ signal.

20. Protection relays are devices that detect abnormal conditions in electrical systems and activate a __________ circuit to disconnect the power source in case of a failure.

9. MEASUREMENT INSTRUMENTS

Voltmeter

The voltmeter consists of a reading display, two measuring terminals, and an internal power source. The terminals are connected to the points between which the potential difference is to be measured, and the voltmeter displays the measurement in units of volts (V). Voltmeters can be analog or digital, and some models can measure voltages up to several kilovolts. Figure 9-1 shows how to connect a voltmeter.

It is important to note that when measuring the potential difference in a circuit, if the voltmeter is not connected in parallel with the component to which the voltage is to be measured, but in series, the measurement could be inaccurate, or the voltmeter could be damaged.

The voltmeter is an essential measurement tool in the electrical and electronic industry, as it allows not only measuring the voltage of electrical components and circuits but also verifying their operation and detecting connection problems or faults. However, it is important to follow the manufacturer's instructions and take safety precautions when using a voltmeter, as working with electricity can be dangerous, as we have emphasized throughout this book.

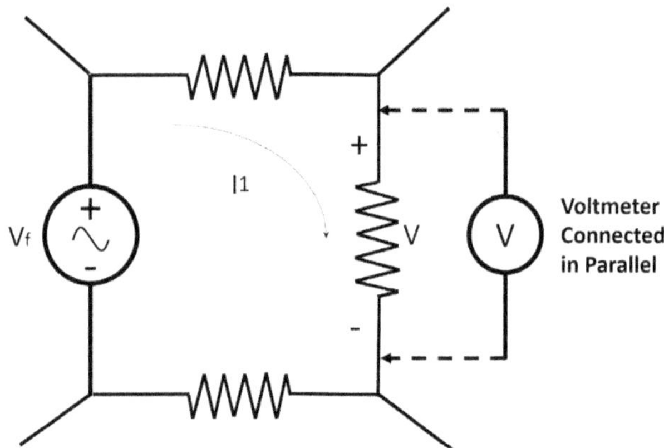

Figure 9-1. How to connect a voltmeter

Ammeter

An ammeter is a measurement instrument used to measure the electric current flowing through a component or circuit. The equipment is connected in series with the component or circuit to be measured, and it measures the electric current flowing through the circuit in amperes. Figure 9-2 shows how to connect an ammeter in a circuit.

The ammeter consists of a reading display, two measuring terminals, and an internal power source. It can be analog or digital, and it can measure electric currents from micro or milliamperes to several amperes.

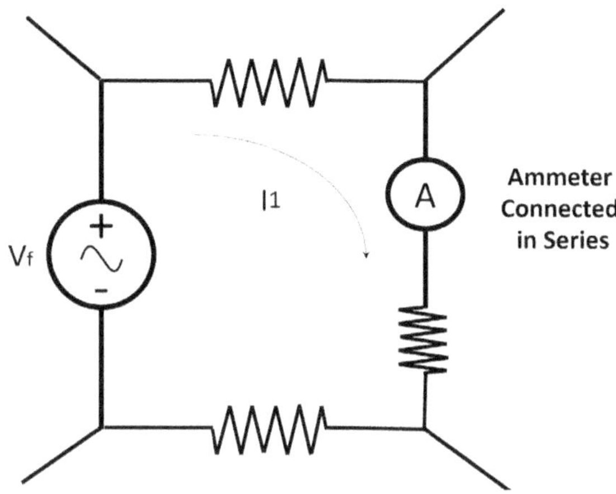

Figure 9-2. How to connect an ammeter

It is important to note that when measuring electric current in a circuit, it is necessary to connect the ammeter in series with the component or circuit to be measured. If it is connected in parallel, the measurement could be inaccurate, or the ammeter could be damaged.

Ohmmeter

An ohmmeter is a measurement instrument used to measure the electrical resistance of a component or electrical circuit. The ohmmeter functions by applying a known electric current to the component and measuring the voltage drop that occurs across it. From this measurement, the electrical resistance of the component can be calculated using Ohm's law (R=V/I). This calculation is performed internally by the device.

The ohmmeter consists of a reading display, two measuring terminals, and an internal power source. The terminals are connected to the ends of the component or circuit to be measured, and the ohmmeter displays the electrical resistance in units of ohms (Ω). It is important to note that to measure the electrical resistance of a component or circuit, it is necessary to disconnect the power supply and ensure that the component has no electric charge. Otherwise, the ohmmeter could be damaged, or the measurement could be inaccurate. Figure 9-3 shows how to measure a resistance with an ohmmeter. Ohmmeters can be analog or digital, and some models can measure resistances up to several megaohms.

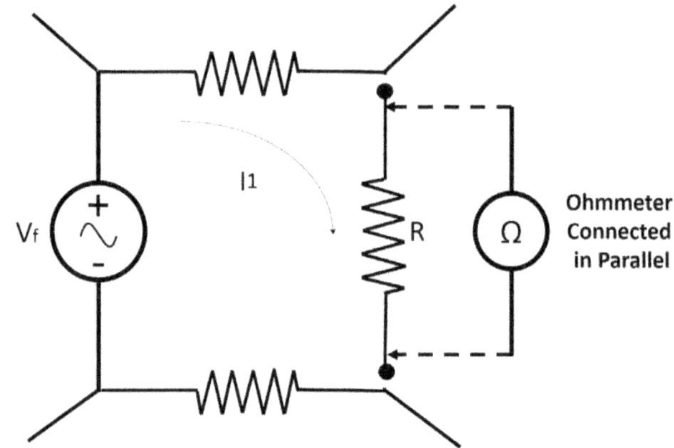

Figure 9-3. How to measure resistance with an ohmmeter

As with the voltmeter and ammeter, the ohmmeter is an essential measurement tool for an electrician, as it allows not only measuring the resistance of electrical components and circuits but also verifying their operation and detecting connection problems or faults. Again, it is important to follow the manufacturer's instructions and take safety precautions when using an ohmmeter, as working with electricity can be dangerous.

Multimeter

A multimeter, also known as a volt-ammeter-ohmmeter, is an electrical measurement tool used to measure current, voltage, and resistance in an electrical circuit. It is essential for technicians, engineers, and enthusiasts who work with electricity.

As with the equipment already seen, the multimeter consists of a digital display (although it can also be analog) that shows the measurements, and two cables with probes that are connected to the electrical circuit to measure current, voltage, or resistance. Most multimeters have a selection knob that allows the user to choose the measurement to be taken, such as volts, amperes, or ohms.

It should be noted that the physical appearance of a multimeter is very similar to that of a voltmeter, an ammeter, and an ohmmeter, both in its analog and digital presentation, with the difference that a multimeter has the ability to perform all measurements (voltage, current, or resistance), one at a time, while the voltmeter only measures voltage, the ammeter only measures current, and the ohmmeter only measures resistance. Figure 9-4 shows the physical appearance of a multimeter. Note the selection knob, which is very important to point to the measurement range to be performed to avoid equipment damage or imprecise measurements.

Clamp Meter

A clamp meter is a type of measuring instrument used to measure electric current flowing through a conductor without interrupting the circuit. The clamp meter consists of a jaw that opens and closes around the conductor and a meter that shows the electric current measurement. See the appearance of this device in Figure 9-5.

Figure 9-4. Appearance of a multimeter

The clamp meter works on the principle of electromagnetic induction. When placed around the conductor, a magnetic field is created that is proportional to the electric current flowing through it. The clamp detects this magnetic field and converts it into an electric current measurement displayed on the meter's screen.

Clamp meters can measure electric currents up to several hundred amperes. Some can also measure voltage and electrical resistance. Clamp meters are widely used in the electrical and electronic industry, for applications such as maintenance and troubleshooting in electrical and electronic systems.

Clamp meters must be calibrated and tested regularly to ensure their accuracy, and all necessary safety precautions must be followed when working with them.

Figure 9-5. Appearance of an amperometric clamp

Wattmeter

A wattmeter is a measuring instrument used to measure the electrical power consumed by an electrical component or circuit. This device consists of a reading display, two measuring terminals, and an internal power source. The terminals are connected in series with the component or circuit to be measured, and the wattmeter measures the electric current and the electric potential difference across the circuit, and then calculates the consumed electrical power.

The wattmeter can be used to measure both active electrical power (measured in watts) and reactive electrical power (measured in volt-amperes reactive or VAR). Reactive electrical power is the amount of electrical energy stored in a circuit and returned to the power supply instead of being consumed by the component or circuit.

Energy Meter

An energy meter, also known as an electric consumption meter or electricity meter, is a device used to measure the amount of electrical energy consumed in a home or building. The energy meter is owned by the electric company and is used to bill customers for their electrical energy consumption.

This equipment measures the electric current flowing through a circuit and multiplies it by the potential difference (voltage) to obtain the consumed electrical power. The consumed electrical power is measured in kilowatt-hours (kWh), which is the amount of electrical energy consumed in one hour at a rate of one kilowatt.

Energy meters have become increasingly sophisticated, with many models offering additional functions such as measuring the quality of electrical energy, remote monitoring, and long-term consumption data storage. Smart energy meters have also been developed, which communicate with the power grid and allow electric companies to adjust energy production according to demand.

It is important to note that the energy meter must be installed and maintained by a qualified professional, and users should take measures to reduce their electrical energy consumption and decrease their environmental impact.

Oscilloscope

An oscilloscope, see its appearance in Figure 9-6, is an electronic measuring instrument used to visualize variable electrical signals over time. It is commonly used in electronics, engineering, physics, and other areas where precise measurements of electrical signals are needed.

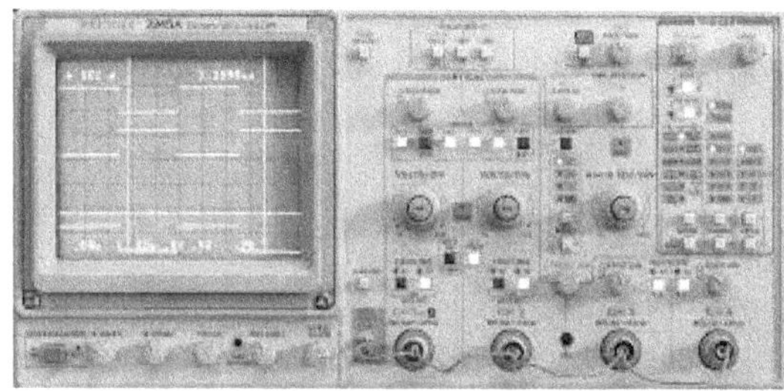

Figure 9-6. Oscilloscope

The oscilloscope works by measuring the electrical voltage in a circuit and then graphically representing it as a waveform on a screen. The screen is typically a cathode-ray tube (CRT) or LCD.

The equipment has two input channels, allowing the simultaneous measurement and visualization of two different electrical signals on the screen. The waveform on the screen can be adjusted using controls for scale, time, and vertical and horizontal displacement.

In addition to visualizing the waveform, the oscilloscope can also be used to measure the amplitude, frequency, and phase of the electrical signal. It can also detect and measure the presence of noise and distortion in the signal.

There are several types of oscilloscopes, including analog,

digital, and mixed-signal oscilloscopes. Analog oscilloscopes use a CRT to visualize the signal, while digital oscilloscopes use an LCD screen and convert the signal into digital form for processing.

It is an essential tool for anyone working with electrical signals and needing to measure and analyze them accurately over time. It is used in a wide range of applications, from the design and testing of electronic circuits to research in physics and other fields related to electronics.

Signal Generator

A signal generator is an electronic measuring instrument used to produce electric signals in a controlled and precise manner. These signals are used for various purposes, such as testing and calibrating electronic equipment, research and development of electronic systems, and teaching and learning signal and system theory.

Signal generators can produce a wide range of waveforms, including sine waves, square waves, triangular waves, pulse waves, and ramp waves. The waveform generated by a signal generator can be adjusted using controls for frequency, amplitude, duty cycle, and other parameters.

In addition to producing simple signals, signal generators can also produce modulated signals, such as AM, FM, and PM. These modulated signals are important in communication and signal processing and are used in a variety of applications, such as radio and television transmission, telephone communication, and data transmission. Also in research and development of electronic systems and teaching signal and system theory at academic institutions.

There are several types of signal generators, from simple analog to more advanced digital ones. Analog signal generators use analog electronic circuits to produce the signal, while digital signal generators use digital circuits and software to generate the signal.

Continuity Tester

A continuity tester is an electronic measuring instrument used to determine if an electrical circuit is closed or interrupted. It is commonly used in electronics, electrical work, and other areas where precise measurements of electrical continuity are needed.

This device works by sending a small electric current through two points of the circuit and measuring the resistance between them. If the resistance is zero, it means the circuit is closed, and the current can flow smoothly. If the resistance is infinite, it means the circuit is interrupted, and the current cannot flow.

The continuity tester can be a simple tool with an LED light or an acoustic signal that indicates whether the circuit is closed or interrupted. It can also be more sophisticated, with an LCD screen and additional functions, such as measuring resistance and capacitance.

It is an essential tool for anyone working with electrical circuits and needing to determine if there is an interruption in the circuit. It is commonly used in the installation and maintenance of electrical and electronic systems, and it is crucial to ensure the safety and proper functioning of electrical systems.

Polarity Tester

A polarity tester is an electronic measuring instrument used to determine the polarity of an electrical circuit. Polarity refers to the direction of the electric current flow in the circuit and is important in many applications, such as the installation and maintenance of electrical and electronic systems.

When connected to a circuit, the polarity tester will indicate whether the polarity is positive or negative. This is important because incorrect cable connections can damage circuit components and endanger people's safety.

Polarity testers can be simple or sophisticated, and some models may include an additional function to test circuit continuity. Simple ones have an LED light or an acoustic signal that indicates whether the polarity is positive or negative, while more sophisticated ones may have an LCD screen and a voltage measurement function.

It is a useful tool for anyone working with electrical circuits and needing to determine the circuit's polarity. It is commonly used in the installation and maintenance of electrical and electronic systems, and it is essential to ensure the safety and proper functioning of electrical systems.

Frequency Meter

A frequency meter is a measuring instrument used to measure the frequency of an electrical or electronic signal. Frequency refers to the number of cycles or vibrations per second that occur in the signal.

The frequency meter works by measuring the signal's period, which is the time it takes for the signal to complete one full cycle. From this measurement, the frequency meter can calculate the signal's frequency as the inverse of the period. Figure 9-7 shows two cycles of a sinusoidal signal.

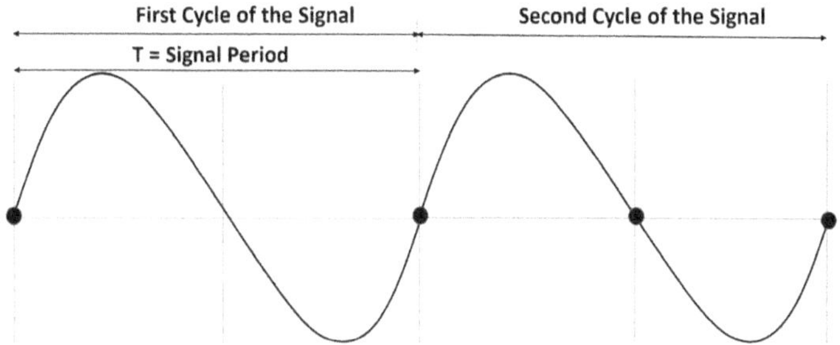

Figure 9-7. Period and cycles of a sinusoidal signal

Frequency meters can be analog or digital and can measure a wide range of frequencies, from a few hertz to several gigahertz. Some models can also measure signal amplitude and other characteristics.

It is used in various applications, such as measuring the frequency of electric current in the power grid, measuring the frequency of radio and television signals, measuring the frequency of musical instruments, and measuring the frequency of pulses in digital systems.

Contact Thermometer

A contact thermometer is a measuring instrument used to measure the temperature of an object or surface by direct contact. Contact thermometers are commonly used in industrial and laboratory applications to measure the temperature of liquids, gases, metals, and other materials.

The contact thermometer works by measuring the change in electrical resistance of a sensor element, such as a thermocouple or a thermistor, which is in direct contact with the surface to be measured. As the temperature of the object changes, so does the sensor's electrical resistance, allowing the contact thermometer to calculate the temperature.

Contact thermometers can be digital or analog and can have different measuring ranges and accuracy levels. Some models may also have additional features, such as data storage or connectivity to a computer for data logging and analysis.

Contact thermometers are used in a variety of applications, from measuring temperature in the food and pharmaceutical industry to measuring temperature in internal combustion engines and welding processes.

Non-Contact Thermometer

A non-contact thermometer, also known as an infrared thermometer, is a measuring instrument used to measure the temperature of an object or surface without the need for physical contact. Instead of measuring temperature through direct contact, non-contact thermometers use the infrared radiation emitted by the object to measure its temperature. Figure 9-8 shows a digital non-contact thermometer.

They work using an infrared sensor that detects the infrared radiation emitted by an object and converts it into a temperature reading. They can measure temperature from a distance, making them ideal for use in situations where direct contact is not possible or desirable, or is dangerous.

Non-contact thermometers can have different measuring ranges and accuracy levels, and some models may have additional features, such as data storage, connectivity to a computer for data logging and analysis, and the ability to measure temperature in different areas of the object or surface.

Figure 9-8. Appearance of a non-contact thermometer

Non-contact thermometers are used in a wide range of applications, from measuring temperature in the food and pharmaceutical industry to measuring temperature in the automotive and aerospace industry. They are also useful for measuring body temperature, making them popular in the medical industry and fever detection.

Ground Potential Meter

A ground potential meter, also known as a ground resistance meter, is a measuring instrument used to measure the electrical resistance of the ground. The main purpose of a ground potential meter is to ensure that electrical installations are safely connected to the ground and that there is a proper grounding connection to prevent dangerous electrical discharges.

This device works by measuring the electrical resistance between a ground electrode and a ground reference, using a known test current. The meter applies current through the ground electrode and measures the generated voltage, allowing it to calculate the electrical resistance of the ground.

Ground potential meters can be portable or stationary, and they can have different measuring ranges and accuracy levels. Some models may also have additional features, such as storing measurement data and performing insulation resistance measurements.

They are commonly used in the electrical and construction industries to ensure that installations are safely grounded. They are also used in applications where measuring ground resistance is important, such as in the telecommunications industry.

Power Factor Meter

A power factor meter, like the one shown in Figure 9-9, is a measuring instrument used to measure the efficiency of electrical energy use in a system and is defined as the ratio of real power (watts) to apparent power (volt-amperes).

A low power factor means that a significant portion of the supplied electrical energy is not used efficiently and is lost as heat. This can result in higher electricity costs and can cause issues in the electrical system, such as increased voltage and overheating of electrical equipment.

The power factor meter works by measuring the active and reactive power (generated by inductances and capacitances) of the electrical system and calculating the power factor. Typically, the meter displays the power factor on a numeric scale or a digital screen.

Figure 9-9. Power factor meter

These devices are used in industrial and commercial electrical systems, especially those with inductive loads, such as electric motors and transformers. These devices can cause electrical current displacement, which affects the power factor. Measuring the power factor is important to ensure optimal energy efficiency and reduce electricity costs.

Characteristics of Measuring Instruments

We have seen that measuring instruments are fundamental tools in electrical and electronic engineering for obtaining precise and reliable measurements of magnitudes. For an instrument to be considered suitable for use in the field of electrical engineering, it must have certain specific characteristics, such as:

1. **Accuracy**: The measuring instrument must be capable of providing precise and accurate measurements of elec-

trical magnitudes.

1. **Sensitivity**: The instrument must be sensitive enough to detect even the smallest variations in the electrical magnitude.

1. **Resolution**: Resolution is the ability of the measuring instrument to distinguish between different levels of electrical magnitude. The higher the resolution, the smaller the variations the instrument can detect.

1. **Range**: The measuring instrument must be able to measure a wide range of electrical magnitudes, from very small to very high values.

1. **Response Time**: The response time of the instrument must be fast enough to measure rapid variations in electrical magnitudes.

1. **Stability**: The instrument must be capable of maintaining constant calibration and accuracy over a long period of time.

1. **User-Friendly**: The measuring instrument must be easy to use and operate for users.

1. **Protection**: The measuring instrument must be designed with adequate protections to prevent damage due to overloads or short circuits.

1. **Portability**: In some cases, it may be important for the measuring instrument to be portable and easy to transport.

1. **Cost**: The measuring instrument must be cost-effective and offer a good balance between cost and quality.

It should be noted that the choice of the appropriate measuring instrument depends on the specific application and the user's needs.

Review Questionnaire

Answer the question:

1. What are the characteristics that measuring instruments must have to ensure their accuracy and reliability in measurement?

2. What is a multimeter, and what is it used for in the field of electronics and electricity?

3. What is the difference between a voltmeter and an ammeter? How are these instruments connected in an electrical circuit?

4. What is an oscilloscope, and what is it used for in the measurement of electrical signals?

5. What is a non-contact thermometer, and how is it used to measure the temperature on various objects and surfaces?

Select the correct option:

6. Which of the following measuring instruments is used to measure electric current in a circuit?

   a. Voltmeter
   b. Ammeter
   c. Ohmmeter
   d. Frequency meter

7. Which measuring instrument is used to measure electrical resistance in a circuit?

   a. Voltmeter
   b. Ammeter
   c. Ohmmeter
   d. Clamp meter

8. Which of the following measuring instruments is used to measure the temperature of objects without the need for physical contacto?

    a. Contact thermometer
    b. Ground potential meter
    c. Non-contact thermometer
    d. Wattmeter

9. Which measuring instrument is used to generate electrical signals of different frequencies and amplitudes?

    a. Signal generator
    b. Oscilloscope
    c. Energy meter
    d. Frequency meter

10. Which of the following measuring instruments is used to determine electrical continuity in a circuit?

    a. Power factor meter
    b. Polarity tester
    c. Continuity tester
    d. Contact thermometer

Answer if the statement is true or false:

11. A voltmeter is used to measure electric current in a circuit.

    a. True
    b. False

12. The non-contact thermometer is a measuring instrument that measures the temperature of objects through physical contact.

    a. True
    b. False

13. The ammeter is connected in parallel in a circuit to measure the electric current flowing through it.

    a. True
    b. False

14. The wattmeter is used to measure the electrical energy consumed by a device in a specific period of time.

    a.   True
    b.   False

Complete with the appropriate word or expression:

15. The multimeter is a measuring instrument used to measure __________, __________, and __________ in an electrical circuit.

16. The voltmeter is used to measure the __________ in an electrical circuit.

17. The clamp meter is a measuring instrument used to measure electrical __________ without interrupting the circuit.

18. The ground potential meter is used to measure the electrical __________ between a ground point and a reference point.

19. The oscilloscope is a measuring instrument used to visualize electrical signals in the form of __________ on a screen.

10. ELECTRICAL SAFETY

Introduction

Electrical safety is a very important topic that must be taken seriously by everyone, from homeowners to electricians and construction workers. Electricity can be dangerous if mishandled, resulting in serious injuries or even death. It is therefore crucial to follow certain electrical safety measures to ensure protection, not only for people but also for properties.

One of the first electrical safety measures to consider is always working with electricity disconnected from the source. Before performing any electrical work, it is necessary to ensure that there is no electric current flowing in the circuit we are going to work on. This can be done by turning off the corresponding switch in the electrical panel or unplugging from the wall outlet. Care must also be taken when working near electrical wires, especially if they are damaged or exposed.

Another important measure is to ensure that electrical wires are in good condition. If they are worn out or damaged, they can cause short circuits or electrical arcs, which can lead to fires or injuries. If any faulty wires are detected, they should be replaced immediately.

It is also essential to consider the type of plug and cable being used. Plugs and cables should be designed to handle the electrical load that will be applied to them. They should not be overloaded, and cables that are too thin for the electrical load they will carry should not be used, as this can cause overheating and fires.

Another safety measure is the use of personal protective equipment (PPE). Electricians and construction workers should wear safety glasses, insulated gloves, and specialized footwear when working with electricity. This protects against the possibility of electrocution or injury in case of an accident.

Lastly, whenever possible, electrical work should be done by a professional electrician. Electricians have the necessary knowledge and experience to ensure that the work is done properly and safely. It is always better to hire a qualified electrician to perform any electrical work, especially for large or complex projects.

Preventing Electrical Risks

Preventing electrical risks is a critical aspect of electrical safety, and it refers to the measures and practices that must be implemented to avoid electrical accidents at work and at home. Electrical risks can be fatal and occur due to factors such as lack of protection, exposure to hazardous electrical elements, and lack of knowledge and proper training.

To prevent electrical risks, safety regulations and electrical standards must be followed, and appropriate safety measures should be established in all electrical installations, whether at home or in the workplace. Here are some electrical risk prevention measures that can be useful:

1. **Inspection and maintenance**: It is essential to perform regular inspections and maintenance of all electrical equipment and systems to identify and resolve issues before electrical accidents occur. Inspections may include reviewing electrical installations, checking the condition of protective equipment, and conducting safety tests.

2. **Training and education**: Workers and homeowners should receive training and education on the proper use and maintenance of electrical equipment and systems. Training may include electrical safety courses, EPP usage training, and conducting drills and practices in emergency situations.

3. **Use of labels and signage**: Labels and signage can help identify hazardous electrical zones and equipment and inform workers and users about potential electrical risks.

Safety Regulations and Standards

Electrical safety regulations and standards are rules and guidelines that must be followed by electricians, construction workers, and homeowners to ensure the safety of people and properties. These regulations and standards establish best practices and minimum requirements for the design, installation, and maintenance of electrical systems.

One of the most well-known and used electrical safety standards is the *National Electrical Code* (NEC) in the United States. The NEC sets the minimum requirements for electrical system installations in buildings and structures. The code includes standards for fire protection, capacity of electrical conductors, installation of switches and outlets, and lighting system installation.

Another important standard is the *International Electrotechnical Commission* (IEC) in Europe, which establishes international standards for electrical safety. The IEC covers a wide range of areas, including protective equipment, medical equipment, renewable energy systems, and industrial control systems.

In the United Kingdom, the *Institution of Engineering and Technology* (IET) publishes a set of rules known as the Wiring Regulations (BS 7671). This regulation sets the standards and requirements for electrical system installations in the country.

In addition to these standards, there are also specific electrical safety regulations for different sectors and applications. For example, the electrical safety regulation for the medical industry deals with equipment used directly on patients, and the regulation for electrical equipment used in hazardous areas, such as the chemical or oil industries, is very specific and rigorous.

In Colombia, electrical safety regulations are regulated by the Ministry of Mines and Energy and are governed by Law 143 of 1994 and the Energy Code.

Law 143 of 1994 establishes the regime for the generation, interconnection, transmission, distribution, and commercialization of electricity in the national territory, grants authorizations, and dictates other provisions in energy matters.

The Energy Code establishes the minimum standards and requirements for the design, construction, installation, operation, and maintenance of electrical installations in Colombia. This code establishes, among other things, the need for an appropriate grounding system, the obligation to carry out periodic inspections, and the need for personal protective equipment to prevent electrical accidents.

Other regulations establish the minimum requirements for the safe operation of electrical installations in Colombia, such as minimum safety for the design, construction, and maintenance of electrical installations, the obligation to have a preventive maintenance plan, and the need for personal protective equipment for workers.

In Colombia, there are also specific regulations for different sectors and applications, such as the regulations for the installation of electrical equipment in hazardous areas (NTE-IEC 60079), the regulations for the installation of electrical systems in buildings (NTC 2050), and the regulations for the installation of photovoltaic systems (NTC 5803).

It is essential to note that in Colombia, as in other countries, companies and professionals working in the electrical sector must comply with electrical safety regulations and standards to ensure the safety of people and properties. Additionally, proper implementation of these regulations and standards can reduce the risks of electrical accidents and improve energy efficiency.

Personal Protective Equipment

Personal protective equipment (PPE) is an important part of

electrical safety in any environment where electrical components are handled. The electrician (engineer, technologist, or technician) must ensure that workers exposed to electrical risks have the appropriate equipment to protect themselves.

PPE for electrical work includes, among others, helmets, safety glasses, gloves, footwear, and insulating suits. This equipment must be carefully selected to ensure that it meets safety requirements and suits the specific needs of each worker..

Electrical safety gloves are a common example of PPE. They must comply with the ANSI/ASTM D120 standard to be considered safe. These gloves are designed to protect workers against electrical shocks and must be tested regularly to ensure that they remain safe over time.

Another example of PPE for electrical work is insulating suits. They are designed to protect workers against static electricity and electrical shocks. The level of protection provided by the suit depends on the voltage to which the worker is exposed. As there are various classification levels, it is important to select a suit that provides the appropriate level of protection for the work to be performed.

It should be noted that PPE is not a complete solution against electrical safety risks. Other procedures, as previously discussed, must also be followed, such as disconnecting power before working with an electrical component, identifying electrical risks in general, and providing training for workers regarding electrical risks and proper use of PPE.

Review Questionnaire

Answer the question:

1. Why is electrical risk prevention important in the workplace?

2. What type of injuries can occur due to exposure to electrical current?

3. What are the main risk factors in the handling of electrical equipment?

4. What safety regulations and standards should be followed in the handling of electrical equipment?

5. What information should an electrical danger warning label contain?

6. What is the objective of regulations regarding the use of personal protective equipment?

7. What personal protective equipment is necessary for handling electrical equipment?

8. How should personal protective equipment be stored and transported?

9. How is the inspection and maintenance of personal protective equipment carried out?

10. What are the consequences of not following safety regulations and standards in the handling of electrical equipment?

THE AUTHOR

ALBEIRO PATIÑO BUILES is an Electrical Engineer with specializations in Literary Hermeneutics and High Management. He holds a Master's degree in Strategic Management, Planning, and Control from the IEE in Spain. He has received numerous awards, including first place in the II National Novel Prize - National Culture Awards from the University of Antioquia (2006) and first place in the First Short Story Contest of the Association of Employees of Banco Industrial Colombiano (1996). His literary publications include: *Historias cruzadas* (tales, 1994), *Bandidos y hackers* (novel, 2007), *Phishing* (novel, 2010), *Construir una novela. Cómo orientarse en el proceso de creación literaria* (essay, 2011), *Intimidación* (novel, 2014), *Galán, crónica de un magnicidio* (novel, 2014), *Las intermitencias del corazón – I. Melancolía y enajenación* (novel, 2016), *La forja de un escritor* (essay, 2017), *Sombras en la Red* (novel, 2019) y *Las intermitencias del corazón - II. Celos y dolor* (novel, 2019)

Collection *Engineering*

www.ingramcontent.com/pod-product-compliance
Lightning Source LLC
LaVergne TN
LVHW020739200726
843506LV00009B/818